MW00669247

The Green Building Materials Manual

Hannah Rae Roth • Meghan Lewis •
Liane Hancock

The Green Building Materials Manual

A Reference to Environmentally Sustainable Initiatives and Evaluation Methods

Hannah Rae Roth
Missouri Gateway Chapter
U.S. Green Building Council, USGBC
Saint Louis, MO, USA

Meghan Lewis
Carbon Leadership Forum
University of Washington
Seattle, WA, USA

Liane Hancock
School of Architecture and Design
College of the Arts
University of Louisiana at Lafayette
Lafayette, LA, USA

ISBN 978-3-030-64887-9 ISBN 978-3-030-64888-6 (eBook)
https://doi.org/10.1007/978-3-030-64888-6

This Springer imprint is published by the registered company Springer Nature Switzerland AG
The registered company address is: Gewerbestrasse 11, 6330 Cham, Switzerland

Acknowledgments

This book represents 12 years of research. Portions of this research were supported at several institutions.

At Washington University in St. Louis our work was initiated by Dean Jerry Sincoff and received continued support from the Sam Fox School of Design & Visual Arts Dean of Architecture Bruce Lindsey. Our Materials Sustainability Standards (MSS) research was sponsored by the Washington University and Brookings Institution Joint Academic Venture Fund with additional grant support from The Skandalaris Center for Entrepreneurial Studies. Co-investigators included: Charles McManis, Thomas and Karole Green Professor of Law Emeritus, School of Law, Washington University in St. Louis; Jorge Contreras, Professor, College of Law, University of Utah; and Dr. Charles Ebinger, Senior Fellow and Director, Energy Security Institute, Brookings Institution. The Undergraduate & Graduate Student Research Team at Washington University in St. Louis included: Sam Fox School of Design & Visual Arts – Meghan Lewis, Sam King, Lydia Slocum, and Jennifer Lee; School of Law – Alex Polley; College of Arts and Sciences – Elizabeth Weingartner. Steve Pentecost aided us with all technology. Special thanks to Enrique Von Rohr, Director of Research and Technology and Senior Lecturer; to the enthusiastic students in Hannah Rae Roth's 11 years of the Arch 434M Materials Research Seminar, and 8 years of Arch 434R Materials Research Seminar: Decoding Sustainability; and to Carmon Colangelo, Dean of the Sam Fox School of Design & Visual Arts, Washington University in St. Louis.

Our Materials Sustainability Standards research resulted in the paper: "Toward a Rational Framework for Sustainable Building Materials Standards." The paper was published in Standards Engineering Society after winning first place in the World Standards Day Paper Competition 2011 [1]. Additionally, the research was published in the paper "Higher Standards for Sustainable Building Materials," in *Nature Climate Change* [2].

On February 24–25, 2011, the research team from Washington University and the Brookings Institution convened a 2-day workshop with a group of widely recognized experts in MSS research and policy. The panel of experts included: Deborah Dunning, the Green Standard; Jennifer Atlee, BuildingGreen; Carole Hetfield,

EPA; Allison Kinn Bennett, EPA; Angie Leigh, EPA; Kevin Gartner, University of New Hampshire; Kristen Ritchie, Gensler; Kyo Suh, University of Minnesota; and Bruce Uhlmann, BASF. The goals of the workshop were to: (a) inform the selection of certifications that would form the basis of our pilot analysis project, (b) inform the range of environmental attributes to be included in the study, and (c) assess contemporary perspectives concerning Materials Sustainability Standards and certifications.

At Louisiana Tech University, our research continued – now focusing upon the development of a sustainable materials database. This research was supported by the LA i6 Proof of Concept Center, funded by the US Economic Development Administration. Co-investigators included: Dr. Sumeet Dua, Associate Vice President, Max P. and Robbie L. Watson Eminent Scholar Chair, Professor, College of Engineering and Science; and Patrick Miller, former Associate Professor and Coordinator Graphic Design, School of Design. Additionally, a total of six students enrolled in undergraduate and graduate programs in architecture, computer science, and graphic design participated. Dr. Dave Norris, Chief Research and Innovation Officer, provided coordination. Special thanks to Dr. Donald P. Kaczvinsky, Dean of the College of Liberal Arts, and Karl Puljak, Director of the School of Design.

The sustainable materials database was designed to provide an infrastructure to connect manufacturers with design professionals. It provided information on material characteristics with in-depth numerical data across performance and sustainability metrics at the scale of big data. This research was presented as a conference paper entitled "Apples to Oranges: Comparing Building Material Data" at the Architecture Research Centers Consortium Conference in 2015, and published in the Perkins & Will Research Journal as "Apples to Oranges: Comparing Building Material Data." [3]

In 2019, this book was commissioned by Springer Nature after the presentation of the paper "Materials Sustainability Standards and Energy" at the IEEE Green Technologies Conference, Lafayette, LA, a conference developed in coordination with the University of Louisiana Lafayette. At University of Louisiana Lafayette, special thanks to Dean Gordon Brooks, College of the Arts; Interim Dean Michael McClure, College of the Arts; Director Kari Smith, School of Architecture and Design; Professor Corey Saft, Undergraduate Architecture Coordinator; Gretchen LeCombe Vanicor, Director of Sustainability; and the Summer Architecture 598L class. While versions of the artwork related to this research have been developed over the years, Carla Ortega Rodriguez provided the final artwork for the figures in this book.

Very special thanks to Michael McCabe, commissioning editor with Springer Nature, who sent us an email asking, "I'm writing to see if you would be interested in exploring the idea of developing a book based on your talk?"

References

1. Contreras, J., Lewis, M., & Roth, H. (2011). Toward a rational framework for sustainable building materials standards. standards engineering, *The Journal Of SES – The Society for Standards Professionals*, 63(5). https://papers.ssrn.com/sol3/papers.cfm?abstract_id=1944523. Accessed 26 Sept 2020.
2. Contreras, J., Lewis, M., & Roth, H. (2012). Higher standards for sustainable building materials. *Nature Climate Change*, *2*, 62–64. https://doi.org/10.1038/nclimate1383. Accessed 26 September 2020.
3. Hancock, L. (2015). Apples to oranges: Comparing building material data. *Perkins + Will Research Journal*, *07.02*, 58–72. http://research.perkinswill.com/articles/apples-to-oranges-comparing-building-materials-data/. Accessed 26 Sept 2020.

Prologue

A Letter from the Authors

The initial research that led to this book started in 2008 at the Sam Fox School of Design & Visual Arts, Washington University in St Louis. It began within the context of educating architecture students about building materials, their characteristics, their use, and significant buildings through history made of these materials. Early in the development of this research, it became obvious that discussing a material's environmental impact – both from the perspective of how it functioned in a building, such as the insulation value of the building envelope, as well as its sourcing and manufacturing – was integral to understanding and describing building materials.

In 2008, when we started focusing on the environmentally sustainable aspects and impacts of building materials, we could not imagine the methods of measuring, data collection, transparency, and innovation in material development that exists in 2020. At the same time, we continue to impatiently push for more innovation and measuring of the impact of building materials.

Over the past 12 years, investigations into the environmental impacts of materials have grown exponentially. The landscape of voluntary environmental standards and certifications that we initially studied has evolved into a much broader landscape of legislation and building codes related to sustainable building materials as well as new or updated standards and certifications.

As we write this book in 2020, the world is in the first 7 months of the COVID-19 pandemic. Protesters are on the streets of many cities, nationally and internationally. Every day's news reveals major revelations and insights into current happenings, politics, and industry's activities. In June, New York Times journalist Lisa Friedman wrote [1]:

> "I can't breathe" was one of the last things George Floyd said before he was killed in police custody in Minneapolis. The words have become a rallying cry for protesters demanding accountability. Increasingly, they've also been taken up by environmental activists seeking

> to address the fact that communities of color carry a disproportionate burden of the country's air pollution and other environmental hazards.

By necessity, we, as authors, had to set a cut-off date for what we could cover in this book. While we aim to deliver an up-to-date guide to the world of sustainable building materials and the standards, certifications, and legislation that govern them, we are not equipped to research and understand the fall-out from these current and unfolding events and news reports. In our view, it is incumbent on every professional and student to stay well informed by reliable news sources and understand the effect of world events on the building industry.

The goals defined by the World Commission on the Environment and Development in the 1980s are still extremely relevant. There has been tremendous progress in understanding the impact of the built environment and our ability to build more resilient, healthy, and equitable places, but we are still far from achieving an equitable, sustainable planet.

Sincerely,
Roth, Lewis, and Hancock

To contact the authors directly please email us at: greenbuildingmaterialsmanual@gmail.com.

Reference

1. Friedman, Lisa (2020) The Environmental Justice Wakeup Call. New York Times Climate Fwd: Newsletter, 17 June 2020. https://www.nytimes.com/2020/06/17/climate/climate-environmental-justice.html. Accessed 19 Sept 2020.

Contents

Chapter 1
Introduction

1.1 Defining Sustainability

1.1.1 World Commission on the Environment and Development

In the 1980s, the Secretary-General of the United Nations (UN) established the World Commission on the Environment and Development – also known as the Brundtland Commission – to propose a long-term strategy and agenda for solving the world's environmental problems [1]. The Commission, led by former Norwegian Prime Minister Gro Harlem Brundtland, brought together leaders and stakeholders from across sectoral, economic, and national divides to understand and define the scope of the environmental challenge facing the globe.

One of the defining characteristics of the Brundtland Commission is its refusal to isolate environmental issues from social and economic challenges, despite the wishes of some in the UN when the scope of the Commission was first discussed in 1982. As stated by Brundtland in the Chairman's Foreword [1]:

> The environment does not exist as a sphere separate from human actions, ambitions, and needs, and attempts to defend it in isolation from human concerns have given the very word "environment" a connotation of naivety in some political circles...the "environment" is where we all live; and "development" is what we all do in attempting to improve our lot within that abode. The two are inseparable...The links between poverty, inequality, and environmental degradation formed a major theme in our analysis and recommendations. What is needed now is a new era of economic growth - growth that is forceful and at the same time socially and environmentally sustainable.

Each stakeholder on the Commission had different environmental concerns and ideas for how each concern should be addressed because sustainability touches the lives of every individual and country differently. As a result, the Brundtland

H. R. Roth et al., *The Green Building Materials Manual*,
https://doi.org/10.1007/978-3-030-64888-6_1

Commission created a human-focused definition of sustainable development that is still widely used today [2]:

> Sustainable development is development that meets the needs of the present without compromising the ability of future generations to meet their own needs.

Unfortunately, there is still a great need for the global cooperation around sustainable development that was called for in the Brundtland Commission. The Commission marks only the beginning of a series of international conversations on the environment and climate change that set goals and strategies for international agreement and action. The establishment of the United Nations Framework Convention on Climate Change (UNFCCC) in 1994 and the resulting annual Conference of the Parties (COP) beginning in 1995 are notable examples of this continued global conversation. The annual COP conventions have resulted in a number of international treaties related to climate change, including the 2015 Paris Climate Agreement.

1.1.2 The Sustainable Development Goals

In 2015, the 2030 Agenda for Sustainable Development was adopted by all UN member states, establishing the 17 Sustainable Development Goals (SDGs) to be reached by 2030 globally. The SDGs continue the mission of the Brundtland Commission to align governments, businesses, communities, and individuals around a set of goals that recognize the interdependence of poverty and environmental destruction. The 17 goals include [3]:

1. *No Poverty*. End poverty in all its forms everywhere.
2. *Zero Hunger*. End hunger, achieve food security and improved nutrition, and promote sustainable agriculture.
3. *Good Health and Well-Being*: Ensure healthy lives and promote well-being for all at all ages.
4. *Quality Education*: Ensure inclusive and equitable quality education and promote lifelong learning opportunities for all.
5. *Gender Equality*: Achieve gender equality and empower all women and girls.
6. *Clean Water and Sanitation*: Ensure availability and sustainable management of water and sanitation for all.
7. *Affordable and Clean Energy*: Ensure access to affordable, reliable, sustainable, and modern energy for all.
8. *Decent Work and Economic Growth*: Promote sustained, inclusive, and sustainable economic growth, full and productive employment, and decent work for all.
9. *Industry, Innovation, and Infrastructure*: Build resilient infrastructure, promote inclusive and sustainable industrialization, and foster innovation.
10. *Reduced Inequalities*: Reduce inequality within and among countries.
11. *Sustainable Cities and Communities*: Make cities and human settlements inclusive, safe, resilient, and sustainable.
12. *Responsible Consumption and Production*: Ensure sustainable consumption and production patterns.
13. *Climate Action*: Take urgent action to combat climate change and its impacts.

14. *Life Below Water*: Conserve and sustainably use the oceans, seas, and marine resources for sustainable development.
15. *Life on Land*: Protect, restore, and promote sustainable use of terrestrial ecosystems, sustainably manage forests, combat desertification, and halt and reverse land degradation and halt biodiversity loss.
16. *Peace, Justice, and Strong Institutions*: Promote peaceful and inclusive societies for sustainable development, provide access to justice for all, and build effective, accountable, and inclusive institutions at all levels.
17. *Partnerships for the Goals*: Strengthen the means of implementation and revitalize the global partnership for sustainable development.

The SDGs were conceived at the United Nations (UN) Conference on Sustainable Development in Rio de Janeiro in 2012 to replace the 2000 Millennium Development Goals (MDGs). The scope of the MDGs was explicitly focused on poverty. According to the UN, the MDGs were instrumental in reducing HIV/AIDS infections and child mortality as well as lifting over a billion people out of extreme poverty [4]. The replacement of the MDGs with the SDGs further recognizes the relationship between poverty, inequality, and environmental degradation that formed a major theme of the Brundtland Commission.

1.1.3 The Hannover Principles

In 1991, William McDonough and Dr. Michael Braungart were commissioned by the City of Hannover, Germany, to create a set of design principles for the Expo 2000 World's Fair [5]. These design principles were developed four years after the Bruntland Commission completed its work. The resulting principles, named *The Hannover Principles: Design for Sustainability,* were presented at the 1992 Earth Summit's World Urban Forum in Rio de Janeiro, Brazil, the following year. While architects had addressed environmental design in many forms throughout the history of architecture, the Hannover Principles brought design into the more recent international conversations around sustainable development that began with the Brundtland Commission.

The Hannover Principles build on the ideas of interdependence put forth by the Brundtland Commission and begin to establish core environmental issues for consideration during design, such as waste and renewable energy. The nine principles are as follows [5]:

1. *Insist on rights of humanity and nature to coexist* in a healthy, supportive, diverse, and sustainable condition.
2. *Recognize interdependence*. The elements of human design interact with and depend upon the natural world, with broad and diverse implications at every scale. Expand design considerations to recognizing even distant effects.
3. *Respect relationships between spirit and matter*. Consider all aspects of human settlement including community, dwelling, industry, and trade in terms of existing and evolving connections between spiritual and material consciousness.
4. *Accept responsibility for the consequences of design* decisions upon human well-being, the viability of natural systems, and their right to coexist.

5. *Create safe objects of long-term value.* Do not burden future generations with requirements for maintenance or vigilant administration of potential danger due to the careless creation of products, processes, or standards.
6. *Eliminate the concept of waste.* Evaluate and optimize the full life cycle of products and processes, to approach the state of natural systems, in which there is no waste.
7. *Rely on natural energy flows.* Human designs should, like the living world, derive their creative forces from perpetual solar income. Incorporate this energy efficiently and safely for responsible use.
8. *Understand the limitations of design.* No human creation lasts forever and design does not solve all problems. Those who create and plan should practice humility in the face of nature. Treat nature as a model and mentor, not as an inconvenience to be evaded or controlled.
9. *Seek constant improvement by the sharing of knowledge.* Encourage direct and open communication between colleagues, patrons, manufacturers, and users to link long-term sustainable considerations with ethical responsibility and re-establish the integral relationship between natural processes and human activity.

The Hannover Principles focused primarily on the design profession's impacts on the environment while acknowledging the dependence of humans on the environment and the impact of design on humans. Today's growing concern about the treatment of workers and communities, in conjunction with the SDGs, has brought broader conversations around equity and sustainability into conversations happening within the building industry.

1.2 Measuring Sustainability

The definition of long-term strategy and goals for sustainable development through international conventions and reports is critical for global alignment and cooperation toward sustainability. However, these goals are more relevant for the scale of a country, industry, or the globe rather than for the scale of a specific building or material. To take action at the smaller scale of individual materials and buildings, more specific indicators and goals for measuring are defined.

Measuring sustainability is key to tracking progress and holding companies, governments, and communities accountable to their goals. Which metrics should be used to measure sustainability is still debated, but there is agreement that measurement is integral to sustainability. This idea was described by Lord Kelvin in 1883 [6]:

> When you can measure what you are speaking about, and express it in numbers, you know something about it; but when you cannot measure it, when you cannot express it in numbers, your knowledge is of a meagre and unsatisfactory kind: it may be the beginning of knowledge, but you have scarcely, in your thoughts, advanced to the stage of science, whatever the matter may be.

Standards and certifications are one way in which these broader sustainability goals are transformed into actionable and measurable targets for a specific building or material. Each standard defines sustainability for a particular material, building, neighborhood, or even city, depending on its target and scope. Standards must be continually updated to keep pace with the shifting strategies and challenges inherent to tackling sustainability. As a result, each updated version of a standard or new rating system provides a snapshot of how the building industry defined sustainability at a given point in time for a specific product or typology.

The US Green Building Council (USGBC) created one of the first attempts to categorize a list of measurable criteria to define sustainability for a building. The USGBC established the Leadership in Energy and Environmental Design (LEED) rating system in 2000 for new construction, with a strong emphasis on operational energy consumption and efficiency. Each LEED criterion defines a sustainability attribute and how that attribute is measured to ensure that data collected from different buildings and different building teams are comparable.

In 2006, another rating system was initiated: the Cascadia Green Building Council launched the Living Building Challenge (LBC), defining 16 requirements to achieve status as a "living building." The International Living Future Institute (ILFI) emerged several years later to manage the standard and certification. USGBC and ILFI are not the only rating systems available for the building industry, but their history provides an excellent means of tracking the industry's change over the past 20 years.

Concurrent with the development of rating systems and standards to measure the sustainability of buildings, a focus on building materials also evolved. The creators of the Hannover Principles, McDonough and Braungart, were also the creators of one of the earliest certifications specific to sustainable materials: Cradle to Cradle (C2C). In 2005, the Cradle to Cradle Certified Products Program was created by MBDC (founded by McDonough and Braungart) with a strong focus on circular material systems and health. Around the same time, NSF International led the development of the first sustainability assessment for carpet, published in 2004 as the NSF/ANSI 140 Sustainability Assessment for Carpet. Since then, NSF International has helped develop a range of other standards for resilient flooring, commercial furnishings fabric, wallcovering products, and single-ply roofing membranes. Many standards and labels were also created in the European Union. In 2015, ILFI published a standard focused on building materials, the Living Product Challenge. Green building certifications have referenced these standards and certifications by defining criteria focused on including certified materials in building design.

The ultimate goal for many sustainability advocates is to have sustainability metrics and requirements codified through federal, state, and local laws and regulations. There is a contrast between the requirements set forth by regulations and the goals set by standards and certifications. This contrast demonstrates the spectrum of action toward sustainability that is present today between requirements codified into regulations to the more comprehensive goals set by certifications and organizations. While standards and certifications often drive initial adoption for sustainability criteria, regulations are key to implementing broader change throughout the industry. Regulations can be particularly important for achieving sustainability goals related to building materials due to the large number of companies and even countries required to conform to a regulation throughout a building material's supply chain.

While regulations may indicate what sustainability metrics and issues have become the well-established baseline for sustainability, new or updated standards and rating systems hint at the challenges that are still mostly unaddressed by the building industry. A few examples of these shifts and new directions are:

- From operational carbon to embodied carbon
- From climate change mitigation to adaptation and resilience
- From human rights to social justice and equity

These are just a few examples of the ongoing shifts in the way the sustainability of building materials is defined. This book endeavors to capture both well-established and newer sustainability metrics, but does not delve into newer, undeveloped areas of sustainability that have yet to be defined by metrics despite their importance. For example, current sustainable design principles and metrics fail to address social justice and equity and are therefore a noticeable gap within the scope of this book.

1.3 The Triple Bottom Line

The survival of our economic systems is both dependent and inextricably linked to environmental and social sustainability. Despite international agreement on this issue through venues like the Brundtland Commission and SDGs, there is still a gap between the definition and measurement of sustainability and the actual action and accountability by businesses required for sustainability goals to be achieved.

In 1994, the British sustainability consultant John Elkington coined the term the Triple Bottom Line to popularize this concept of the connection between environmental, social, and economic health in the context of a company's profit and purpose [7]. While the bottom line refers to the total remaining profit after the net of the gains and losses for a business has been totaled, the triple bottom line aims to tally social and environmental gains and losses alongside economic ones.

The application of the triple bottom line often happens in the form of directly converting environmental and social impacts of a company's actions and products into a dollar amount. For example, gains from positive marketing related to sustainable products or losses from increased price of materials due to resource scarcity can be estimated and projected alongside profit for a company.

However, many argue that converting environmental and social impacts into a dollar amount does not adequately address the significance of their impacts. Eric Davidson addressed this issue in 2001 in his book *You Can't Eat GNP* [8]:

> Equating the ecological systems with economic systems strictly on a dollar-for-dollar basis misses the point that the economic system cannot exist without the ecological system. The economic pyramid is always contained within the larger ecological pyramid. The ecological system must remain healthy at all scales — local, regional, and global—if the economic system is to survive.

Elkington and Davidson both aimed to provoke a broader conversation and shift away from companies defining their success solely through financial profit. Instead, they seek a more holistic definition of success that accounts for planetary and human health. Even if environmental and social impacts could be perfectly accounted for by one company, the direct conversion of financial gains and losses does not usually take into account the interdependencies between one company and its context. The triple bottom line eliminates all 'externalities' which are 'free' to an individual company but literally costly to individuals, communities, and governments. Later, society sees these costs through higher expense, lack of availability, and at a global scale, damage to the environment and communities.

Two examples of the challenges of the literal application of the triple bottom line by calculating financial gains and losses related to sustainability are time and geographic location. Both issues are particularly acute in the production of building materials.

First, as identified by the Brundtland Commission, some sustainability goals and challenges are long-term. The decisions made by individuals, businesses, or governments today may not have an impact until the next generation. Other sustainable initiatives have a more immediate effect. For example, operational efficiency strategies such as increased energy and water efficiency and reduced waste result in savings for a building owner within a year or a number of years. As another example, if a company purchases reusable cups rather than providing single-use plastic cups for its employees, the company will save money on purchasing after a relatively short period of time. The same can be said for durable building materials: if a building owner purchases more durable materials, they will save money as well as resources by avoiding the need to replace the materials for a longer period of time.

However, many losses related to sustainable building materials require more time for their impact to become visible. This requires that consumers and owners think beyond their own building to the impact on future projects: even if the price of an unsustainable material is low now, that price could double or triple over time due to resource scarcity and environmental damage across the global supply chain. For example, tropical hardwoods have been depleted to the point of scarcity. A generation ago, a tropical hardwood could be purchased so cheaply that some were used for shipping pallets. The same wood today is so precious and costly that it is only used for luxury finishes and furnishings.

Second, financial losses related to sustainability are spread across a wide geographic location. This means they may not have an immediate or direct impact on the source of harm. For example, the impacts of climate change are felt first and more acutely in regions with a greater likelihood of natural disasters, though ultimately every part of the world is impacted.

The geographic challenge of calculating gains and losses related to sustainability is particularly applicable for building materials due to their global supply chains. The depletion of environmental resources, destruction of human and wildlife communities, unjust working conditions, and social inequalities are often hidden from consumers across the globe. While general hazards may be known, it is easy for consumers to assume that the products they are purchasing are not associated with a particular concern, such as child labor.

The unequal geographic distribution of impacts caused by inaction toward sustainability goals, as well as the history of which countries are responsible for the most environmental damage, is addressed explicitly by many of the international commissions and conventions discussed earlier in this chapter. The UNFCCC specifically "puts the onus on developed countries to lead the way," as described in their summary of the Convention [9]:

> The idea is that, as they are the source of most past and current greenhouse gas emissions, industrialized countries are expected to do the most to cut emissions on home ground. They are called Annex I countries and belong to the Organization for Economic Cooperation and Development (OECD).

While the responsibility for the current state of sustainability varies widely across the globe, achieving sustainability goals requires individuals, communities, companies, and governments to look beyond their borders and their generation. Unfortunately, there is not much imagination required to see the impacts of unsustainable buildings and building materials: the social, environmental, and economic health of our planet is in rapid decline. It is time to act.

References

1. World Commission on Environment and Development. (1987). *Our common future, chairman's foreword*, p. 6. https://sustainabledevelopment.un.org/content/documents/5987our-common-future.pdf. Accessed 13 Sept 2020.
2. International Institute for Sustainable Development. (2020). *Sustainable development*. https://www.iisd.org/about-iisd/sustainable-development. Accessed 7 Sept 2020.
3. United Nations Department of Economic and Social Affairs. (2020). *The 17 goals*. https://sdgs.un.org/goals. Accessed 7 Sept 2020.
4. United Nations Development Programme. (2020). *Background of the sustainable development goals*. https://www.undp.org/content/undp/en/home/sustainable-development-goals/background.html. Accessed 7 Sept 2020.
5. Braungart, Michael, William McDonough Architects. (1982). *The Hannover principles*. EXPO 2000, City of Hannover, Germany. https://mcdonough.com/writings/the-hannover-principles/. Accessed 13 Sept 2020.
6. Oxford University Press. (2020). *Oxford reference*. https://www.oxfordreference.com/view/10.1093/acref/9780191826719.001.0001/q-oro-ed4-00006236. Accessed 13 Sept 2020.
7. Elkington, J. (2018). 25 Years ago I coined the phrase "triple bottom line." here's why it's time to rethink it. *Harvard Business Review*, *25*, 2–5. https://hbr.org/2018/06/25-years-ago-i-coined-the-phrase-triple-bottom-line-heres-why-im-giving-up-on-it. Accessed 13 Sept 2020.
8. Davidson, E. (2001). *You can't eat GNP: Economics as if ecology mattered* (p. 59). Cambridge, MA: Perseus Publishing.
9. United Nations. (2020). *What is the United Nations framework convention on climate change*? https://unfccc.int/process-and-meetings/the-convention/what-is-the-united-nations-framework-convention-on-climate-change. Accessed 29 Sept 2020.

Chapter 2
Describing Building Materials and Products

2.1 Introduction

This book outlines environmentally sustainable initiatives and evaluation methods for green building materials focusing upon six sustainability impact categories (Fig. 2.1).

Before concentrating on sustainability across the building material industry, it is important to define a broader organizational structure to understand how material makeup, processing, and purchasing relate to sustainable initiatives (Fig. 2.2).

- What are building materials and products?
- How are they used?
- How are they made and where?
- How are they purchased? Who installs them?

It is easy to begin to list the materials and products that make up a building:

- Lumber
- Plywood
- Window
- Countertop
- Roof
- Sink
- Furnace

But what underlying structure and organization describe these elements, and can that insight reveal how and when different environmental initiatives apply to elements within the building industry?

H. R. Roth et al., *The Green Building Materials Manual*,
https://doi.org/10.1007/978-3-030-64888-6_2

SUSTAINABILITY IMPACT CATEGORIES

Fig. 2.1 Sustainability impact categories

MATERIAL INGREDIENTS

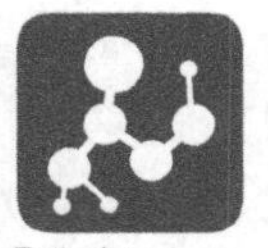

Composite Or Hybrid Materials Include More Than One Ingredient

APPLICATION

PRODUCTS

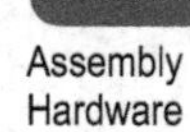

UNIT OF MEASURE

Fig. 2.2 Describing building materials and products through their ingredient makeup, use, product characteristics, and unit of measure

2.2 What Is a Building Material?

A building material is a basic element of construction. At its most simple, it is made from raw materials – this may be cotton, or iron ore, or silica sand (Fig. 2.3). Raw materials become building or construction materials after processing.

Materials fall into categories (Table 2.1), for example:

- Metal and minerals
- Glass and ceramics
- Stone, masonry, cement, concrete
- Plant-based and animal-based
- Petroleum-based plastics
- Composite and hybrid materials with more than one ingredient

This book groups material categories based on their primary ingredients and processes. For example, glass and ceramic are a category because they are both composed of tiny particles that are heated at extremely high temperatures, melting the particles together to create the building material.

The material makeup of a building material or product directly relates to its environmental impacts. Environmental impacts can be both negative and positive, resulting in tradeoffs.

2.3 How Are Building Materials and Products Used?

Building materials and products have many applications and uses. These include:

- Landscaping
- Exterior cladding and window systems
- Interior finishes
- Structure
- Roofing
- Wall assembly materials like insulation and vapor barriers
- Furniture, fixtures, and equipment including materials and equipment associated with mechanical, electrical, and plumbing systems

The specific need for a material or product is often directly related to the material makeup or ingredients. As established previously, the material makeup relates directly to sustainable impacts. For example, most roofing is asphaltic, tar, or rubber. These materials excel at being waterproof, keeping moisture out of a building. However, their imperviousness directly relates to the sealant characteristics of petroleum-based materials and their associated environmental tradeoffs.

A building material's use also dictates how much occupants will come into contact with a product over time. For issues of human health and toxicity, this has a large impact. For example, an occupant is more likely to have direct contact with furniture than a piece of lumber inside of a wall cavity. As a result, there is a greater

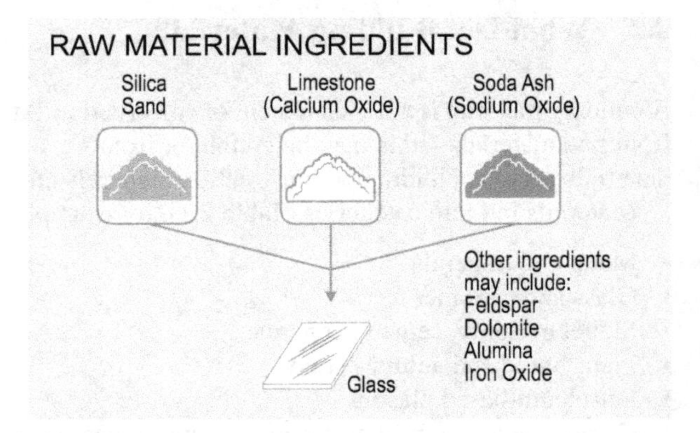

Fig. 2.3 Raw material ingredients for glass [1]

Table 2.1 Examples from material categories

Material Category	Example Opportunities	Example Risks
Metal	*Long product life*: a durable material that lasts decades *Recyclability*: an efficient market for recycling greatly reduces the need for mining	*Large carbon footprint*: smelting and alloy processes require intense amounts of energy *Air and water emissions*: extraction of metal can result in toxins released into the air and adjacent bodies of water *Worker safety*: mining poses large health and safety risks to workers, and international labor law is not always followed
Glass	*Local common materials*: material ingredients are common and easily sourced *Recyclability*: glass can be recycled *Human health*: glass contributes to daylighting, important to human health *Energy efficiency*: advances in technology after the 1970s energy crisis have led to highly engineered energy-efficient window systems	*Large carbon footprint*: uses a tremendous amount of energy to melt and form glass *Air emissions*: heating process can produce toxic emissions
Concrete	*Local common materials*: material ingredients are common and easily sourced *Long product life*: a durable material that lasts decades *Passive heating and cooling*: mass contributes to passive heating and cooling *Use of waste materials*: fly ash, a byproduct of coal burning, has been added as an environmental alternative to powdered cement (with the decrease in coal-powered plants, fly ash is no longer as prevalent)	*Heat island effect*: when used in cities, concrete can contribute to heat island effect *Air emissions*: manufacturing produces hazardous particles that can become airborne *Large carbon footprint*: cement production process is energy intensive and releases large amounts of greenhouse gas emissions

(continued)

Table 2.1 (continued)

Material Category	Example Opportunities	Example Risks
Plant-based	*Local common materials* *Low toxicity* *Negligible carbon footprint* *Biodegradable*	*Toxicity*: not all sealers, binders, and adhesives are non-toxic *Worker safety*: harvesting and manufacturing operations may not adhere to international labor laws
Leather	*Natural material*	*Large carbon footprint*: raising cattle creates a large carbon footprint *Toxicity*: processing uses toxic chemicals *Water use*: processing requires a large volume of water
Petroleum-based (plastics)	*Recycled ingredients*: chemical feedstocks (ingredients) are recycled plastic *Recyclable*	*Carbon footprint*: petroleum-based *Air and water emissions*: may release VOCs initially or toxins into water as they break down

probability that the occupant will be affected by off-gassing or coatings on the furniture versus any toxic sealant the lumber may off-gas. For this reason, discussion of environmental health attributes focuses mainly on interior finishes. However, just because a product is not toxic to a person during building occupation does not guarantee it will cause no health impacts for workers during manufacturing, assembly, construction, or demolition.

2.3.1 Performance

In many cases, a treatment or coating is added after manufacturing to a building material or product to enhance its performance. Materials may be treated to (Fig. 2.4):

- Change their acoustic properties
- Respond to moisture
- Prevent fire and smoke
- Prevent the effects of sunlight
- Provide durability and resistance to abrasion, chemicals, viruses, or bacteria

Unfortunately, many of these treatments or coatings have negative environmental and health impacts due to toxicity of the ingredients or to the release of chemicals known as volatile organic compounds (VOCs) into the air after installation. For example, antimicrobial and fire retardant coatings have been found to be toxic, specifically cancer causing (see Chap. 8 for further discussion). As the industry learns more about the health impacts of these coatings, manufacturers, specifiers, and

MATERIAL PERFORMANCE

Moisture

Absorbent
Porous
Wicking
Treated/Sealed
Water Resistant
Waterproof
Impervious
Moisture Resistant

Acoustic

Sound Reflecting
Sound Diffusing
Sound Absorbing
Sound Deadening

Fire

Fireproof
Fire Retardant
Fire Resistant
Flame Retardant
Flame Resistant
Heat Resistant
Smoke Resistant
Self Extinguishing
Fire Suppression
Class A, Class B, Class C
Class 1, Class 2

UV (Sun)

UV Resistant
Fade Resistant
Color Fast
Lightfastness

Durability & Resistance

SURFACE RESISTANCE
Bulletproof
Puncture Resistant
Impact Resistant
Scratch Resistant
Sag Resistant
Wear Resistant
Stain Resistant
Soil Resistant

ANTISTATIC
Anti-Static
Static Control

FRICTION
Skip Resistant
Skid Resistant
Slip Resistant

CHEMICAL
Anti-Corrosive
Chemical Resistant
Bleach Resistant
Acid Resistant

BACTERIA
Antibacterial
Bacteria Resistant
Bacteriostatic
Bactericidal
Non-Porous
Mildew/Mold Resistant
Antimicrobial
Anti-Allergenic

Fig. 2.4 Describing building materials and products through their performance characteristics

designers are searching for ways to achieve similar performance characteristics through different design strategies.

2.4 How Are Building Materials and Products Made?

Building materials and products are made through:

- Processing
- Manufacture
- Fabrication

Processing entails refining or altering a material through chemistry or mechanical means. For example, gravel must be cleaned and sorted before it becomes aggregate for concrete.

A manufacturing process makes a material conform to a specific size, shape, design, and grade of material quality or strength. For instance, a tree trunk goes through a manufacturing process to become dimensional lumber. Across the United States, a wood 2 × 4 is always the same size and length. It is also graded for quality based on a nationwide standard.

A fabrication process is often multistep. These processes involve assembling manufactured items, or specifically tailoring a product size or design to its intended installation site. A countertop is fabricated when it is made to fit into a corner at a laboratory. The fabrication process ensures that the correct exposed edges and corners are finished and smooth. Within the building industry, the terms manufacturer and fabricator are sometimes interchanged.

Materials can also be organized by their specificity [2]:

- Commodities: Often these are processed raw materials such as cotton or iron ore. They may also be items like nails that can be bought in bulk and are indistinguishable.
- Materials and Products: These are standardized items that can be used interchangeably. They are made by a specific manufacturer or fabricator.
- Proprietary or Name Brand Products: These are products that add value by distinguishing themselves through design, durability, branding, or cost. These products may carry a patent and often have a registered or trademarked name.

Whether an item is a commodity, building material, or name brand product may have bearing on the amount of information one can find on its environmental impacts. There are two main ways this data is affected:

- Chain of custody across supply chain
- Trade secrets

From a simple processed raw material to a complex assembled product, each stage of production introduces new materials or items. These new materials and items come from processors, manufacturers, and fabricators. These different companies create a network or web called the supply chain. An account of the history of who supplied processing and parts is called chain of custody. Often keeping track of this history is difficult.

2.4.1 *Chain of Custody*

Even for relatively simple items, the manufacturing process from resource extraction to distribution can involve many companies. The more complex the building material, the more complex the supply chain. For example, the supply chain for lumber might look like this (Fig. 2.5):

- A tree farm owns the land growing trees.
- A lumber company contracts to cut those trees, stripping them to just the tree trunks.
- A transportation company hauls the trees to the processing plant.
- The lumber mill processes the wood into lumber.
- The lumber is then sold to a distributor, who then sells to either a building materials supplier or a retailer.
- To get from the manufacturing processing plant to the distributor and retailer involves additional transport companies.
- Back at the original site, a company clears the land of the unusable portions of the trees – often turning this excess into wood chips for sale to a company that purchases them in bulk and then sells them for companion uses.

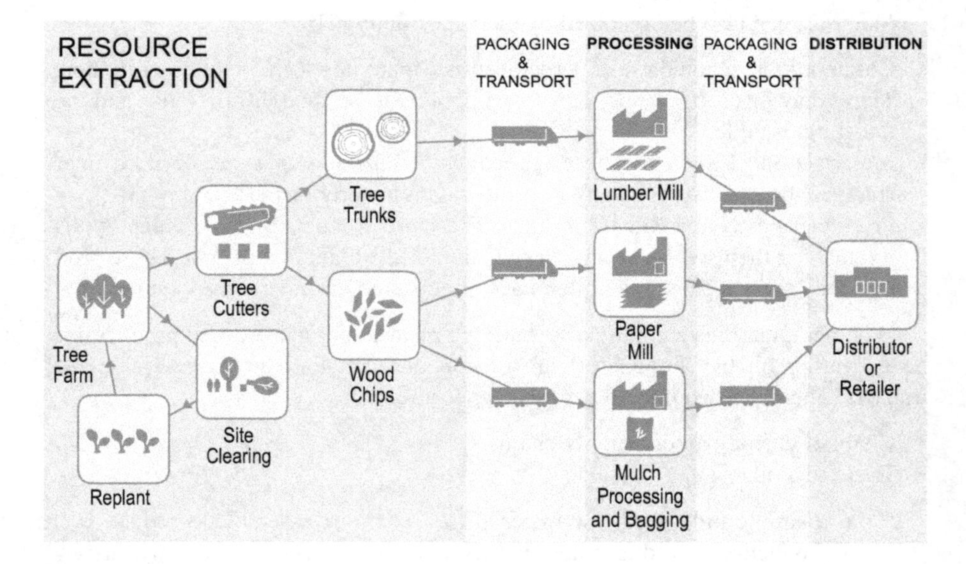

Fig. 2.5 Example diagram for a lumber supply chain

- Finally the original tree farm contracts to replant with new trees, buying those trees from a grower or nursery.

Building materials and products are originally sourced from raw materials. Many of these raw materials are purchased from a commodity market where materials are sorted by grade and bought in bulk by price. Within the commodities market, raw materials from different suppliers are not distinguished. For example, a purchaser may buy cotton from Texas, Louisiana, or Georgia. The purchaser buys by volume or weight, and sources may be mixed to meet an order. Because exchange across the supply chain is based on cost, volume, and weight, often the history of involved companies is lost as material is exchanged from one supplier to the next.

With this mixing of sources, it becomes difficult to record the specifics of:

- How an individual producer extracts or gathers the resource material
- How workers are treated
- The amount of energy or water used
- The release of toxicity that occurs during initial processing

Beyond regional, state, and countrywide laws and regulations, commodities producers are free to behave as they like, as long as their materials meet established standards for quality, dimensioning, and testing.

Many companies choose to brand their building materials or products. For example, plywood may have a "Boise Cascade" label. Name recognition can be associated with quality or cost. Increasingly it is also associated with commitment to sustainable initiatives. In such cases, the company may seek to certify that they have traced all the participants in their supply chain to verify each member's behavior in relation to sustainable initiatives. This includes keeping a history of commodities suppliers and their practices as a chain of custody document.

A fabricated product is often more complex and made of more than one component. While tracking sustainability initiatives at the fabrication facility is achievable, following the suppliers of components and their sustainability initiatives becomes difficult simply because of the complexity of the supply chain. However, fabricators who align their brand with sustainability often seek to manage this process.

2.4.2 Trade Secrets

Proprietary or name brand products distinguish themselves by carrying a patent or having trade secrets. Trade secrets include proprietary manufacturing processes, or secret ingredients or recipes. This information is considered intellectual property owned by a company.

A manufacturer may be hesitant to reveal information about its product data. For example, a company may not want to disclose the chemical ingredients in its raw material supply (also known as feedstock). This may be for a number of reasons:

- By revealing chemical feedstock and other materials, a manufacturer may become vulnerable to competition.
- A manufacturer may not know the information if it has been lost along the supply chain.
- A manufacturer may buy from a supplier who is protecting its own trade secrets – so even if the manufacturer wanted to publicize its product data, it would be limited by what its supplier chose to disclose.
- Disclosing data also indicates associated toxicity, energy use, water use, and possible emissions that a company may not want to reveal.

A manufacturer who aligns its branding with environmental initiatives must decide whether disclosing their product makeup to verify their commitment to sustainability outweighs the benefits of holding trade secrets. Certifications and declarations provide a solution that can solve this issue. For a certification, a manufacturer can reveal data to a certifier and earn a sustainability certification without necessarily making the data public.

2.5 Why Is the Size or Unit of a Building Material or Product Important?

Similar building materials and products may use very different methods of processing or manufacturing. Fundamental to comparing items is agreement upon a common unit of measure for analysis. For example, to compare carpet manufactured by different companies, analysis may be based upon a linear yard of material. This allows for a consistent assessment of the resources the products use and the pollution the manufacturing process produces.

Materials may be measured in terms of volume or area. Each materials sector, such as lumber or carpeting, has its own standards for measurement. Comparison between materials becomes complicated if some items are measured on the imperial system while others, often produced by international manufacturers, are measured on the metric system. Additionally, the area or size of a material may be measured in imperial units, feet and inches, or yards, while the thickness may be measured in metric units, such as millimeters, or in sector specific measurements such as gauge. If a specifier or designer attempts to compare materials with different measuring systems, it falls upon them to calculate the conversion factors. This becomes more difficult if a specifier or designer is attempting to compare materials across sectors. For example, it is challenging to compare the sustainable impacts of ceramic tile flooring, measured in area, with an epoxy paint surface, measured in volume, with both providing a similar use – a cleanable walkable surface.

Complications also exist in the measure of toxins, emissions, and energy. The measurement of emissions and toxins (discussed in Chaps. 7 and 8) is controlled by regulation resulting in common units of measure. However, in this case, confusion is generated by the threshold at which they are reported. For example, a report of formaldehyde may be at 100 or 1000 parts per million. Energy use assessment becomes challenging when some components are made in countries that adhere to the metric system while other components are made in countries that follow the imperial system. This makes the calculation of total energy consumption difficult.

2.6 Where Are Building Materials and Products Made?

Country of origin has a large influence on the environmental impact of a product. Two examples include embodied energy and regulations related to labor and human rights. While one might think about how shipping distances between countries impact embodied energy, the electrical grid of the origin country can be even more important. For example, if a chair is manufactured in Poland and then shipped to Sweden for final assembly, the embodied energy associated with the predominantly coal power grid in Poland would be significantly larger than the energy used during assembly in Sweden. For labor and human rights, when US manufacturers have facilities, subsidiaries, or suppliers in other countries, they may or may not require that those manufacturing facilities comply with US labor regulations.

Companies that wish to sell across international markets follow the national laws and regulations of their own country of origin and may elect to comply with another country's regulations, especially if the building material or product is sold in that country. In particular, a company may meet the requirements of sustainability standards that are recognized in a specific country, so that the product can bear a certification mark or ecolabel that buyers or consumers recognize.

Within the United States, both federal and state regulations govern resource extraction, recycling of materials, toxicity, energy use, water use, and social accountability. A building material or product made in the United States adheres to these regulations. It may also voluntarily comply with one or more standards to achieve a certification or ecolabel.

Some companies seek a "Made in the USA" label as a way to increase brand identity and support local sourcing. The "Made in the USA" label is governed by the Federal Trade Commission. In order to display this mark, a material or product must be made "all or virtually all" in the United States. An alternate label, "Assembled in the USA," denotes that "principle assembly" of foreign components occurs within the United States. In the case of "Made in the USA," building materials and products adhere to federal laws and regulations, whereas "Assembled in the USA" must only meet federal laws and regulations during the portion of assembly that occurs in the United States.

Often the "virtually all" terminology references the *de minimis* rule of up to 7% of material by cost originating outside of the United States. Another consideration is "substantial transformation." In this case, a material or product is labeled as originating in a country where it was transformed by a process giving it a new name, character, or use. For example, a cotton fiber from Turkey may be made into a cotton fabric in Italy. Because it was transformed from a fiber into a fabric in Italy, it can have the label "Made in Italy." A country's Customs Office often makes the final decision on such classifications [3, 4].

2.7 How Are Building Materials and Products Selected?

Building materials and products loosely separate into two categories:

- Commercial building materials and products
- Consumer, retail, or residential building materials and products

These categories are established based on:

- Application or use including building code requirements
- Volume of purchase
- Quality of grade and durability
- Particularity of design and style
- Cost

Commercial building materials and products are typically specified by a designer or facility manager and purchased by a contractor or owner. A design document package is composed of construction drawings and a specification manual. The specifications define:

- The building materials and products included in the project
- Approved methods of installation
- Required testing for performance

Designers and facility managers learn about building materials and products through visits to showrooms, exhibitors at product shows or conferences, product representatives, and product literature. To ensure a contractor or client purchases a specific building material or product, the designer or facility manager uses a proprietary specification, which denotes manufacturer, fabricator, brand, or item number. In situations where a project is publicly bid, a performance specification is used. This specification outlines performance characteristics of building materials or products, and any product meeting those characteristics can be used. This allows flexibility in selecting the lowest priced item. Both proprietary and performance specifications are numbered according to the Construction Specifications Institute (CSI) MasterFormat, encompassing 16 categories ranging from material selection to construction methods. If an owner feels strongly about sustainable initiatives, a designer or facility manager can:

- Write a performance specification that includes sustainability performance characteristics or standards, such as meeting a certain testing standard
- Require that a material or product be certified to a specific sustainability standard within a performance specification

- Require that a material or product be listed in an environmentally preferred purchasing program

Individual consumers typically purchase retail or residential-grade building materials and products. Consumers select and purchase items through:

- A hardware store: These are materials available on the shelves, but in some cases, materials can be ordered. Typically, a hardware store purchases from a select group of distributors or manufacturers. When the hardware store negotiates contracts, those distributors or manufacturers may change and be replaced.
- Through an online retailer: Items may be individually ordered. Some retailers specialize in discounted overstock or discontinued items, especially furniture, fixtures, and equipment.
- At a showroom a consumer orders a building material or product: In the case of a stone countertop, this is a direct order and sale to a fabricator; in the case of a plumbing supply showroom, the showroom represents a selection of brands.

Typically, on both commercial and residential-grade materials, sustainable certification is denoted with an ecolabel affixed to the packaging or item. Some standards and certifications are geared to cover both the commercial and retail market, while others are more tailored to one or the other. Programs like "Energy Star" and "WaterSense" (both run by the EPA) have high name recognition. By contrast certifications for specific sectors, such as "LEVEL" for upholstery, market to designers and specifiers and are less recognized by consumers. The task for a specifier or designer becomes educating a client or owner on the sustainable initiatives met by a certified building material or product. In a retail situation, a consumer must educate themselves as to what a particular ecolabel or certification means.

2.7.1 Who Represents Building Materials and Products?

Many sectors of the building materials industry have developed trade groups or organizations. These organizations serve to represent the sectors to governmental entities, to advertise products, and to provide a forum to discuss advances. A selection of trade groups and organizations includes:

- Carpet industry
- Furniture industry
- Lumber/wood industry
- Plastics industry
- Steel industry/aluminum industry
- Roofing industry
- Upholstery industry

Several of these industries have initiated sustainability standards and certifications. A standard typically measures one or more environmental issues specific to the industry or sector. For example, NSF/ANSI 140 is a certification for carpets that evaluates a range of sustainable impacts and initiatives. It was created by the carpet industry. Sometimes a trade organization leads the sector with its standard for sustainability. In other cases, competing standards or certifications for a specific sector may be developed by non-profit sustainability organizations. Further detail on these standards is presented in Chap. 10.

References

1. Pilkington NSG Group. (2020). *Raw materials*. https://www.pilkington.com/en/global/about/education/the-float-process/raw-materials. Accessed 2 Aug 2020.
2. Hofstrand, D. (2019). *Commodities versus Differentiated products*. Ag Decision Maker, Iowa State University. https://www.extension.iastate.edu/agdm/wholefarm/html/c5-203.html. Accessed 2 Aug 2020.
3. Federal Trade Commission. (1998). *Complying with the made in the USA Standard.* https://www.ftc.gov/system/files/documents/plain-language/bus03-complying-made-usa-standard.pdf. Accessed 6 Aug 2020.
4. Gallagher Logistics and Transport. (2018). *Customs 101: Country of origin*. https://www.gallaghertransport.com/country-of-origin/. Accessed 6 Aug 2020.

Chapter 3
Decoding the Ways to Measure Sustainability and Life Cycle Thinking

3.1 Decoding the Ways to Measure Sustainability

The ways in which building materials and products reduce negative environmental and social effects are impacted by many factors. Owners, designers, specifiers, and consumers request that manufacturers commit to a wide variety of environmental initiatives. At the same time, manufacturers must weigh how those environmental initiatives bring value to their business goals and manufacturing processes. Additionally, years of legislation and a range of sustainability standards have established rules by which environmental and social impacts are measured and controlled. But these rules may or may not align with both consumers' and manufacturers' current concerns. The result is a complex landscape of decision-making and tradeoffs.

Because the field is changing quickly and there are so many things to consider, decoding the metrics and strategies used to measure the sustainability of building materials can be intimidating. The goal of the next six chapters is to aid the reader in understanding the core concepts behind these metrics and strategies by organizing them into six categories (Fig. 3.1):

- Resource use
- Energy
- Water
- Emissions
- Toxicity and human health
- Social accountability

H. R. Roth et al., *The Green Building Materials Manual*,
https://doi.org/10.1007/978-3-030-64888-6_3

SUSTAINABILITY INITIATIVES & ENVIRONMENTAL IMPACTS

Resource Use

Non-Renewable Resources:
Ecosystem Impacts
Local Sourcing
Quarrying, Mines, & Wells
Conflict Resources
Renewable Resources:
Biobased
Rapidly Renewable
Wood Sourcing
Carbon Sequestering

Circular Feedstocks:
Reused/Reclaimed Content
Recycled Content
Dematerialization
Waste Recovery and Circularity:
Biodegradability/Compostibility
Recyclability
Designed for Disassembly
Waste Recovery/Take Back Program
Reduction in Packaging

Energy

Energy Source: Energy Fuel vs Flow
Renewable Energy: Grid, On Site, & Exported
Energy Savings and Efficiency:
Energy Consumption

Audits
Energy Efficiency
Excess and Potential Energy:
Energy Recovery
Energy from Waste
Bio-energy from Waste

Water

Extraction: Water Sourcing
Body of Water Protection
Fresh Water vs. Potable Water
Manufacturing and Use:
Volume Reduction
Water Audits
Water Efficient Tools, Equipment, & Appliances
Net Zero Water Use

Origin: Ecology, Social Accountability
Embodied Water
Water Footprint
Water Stewardship
Worker Water Supply
End of Life:
Water Recycling
Waste Water Quality
Antidegradation Requirements
Water Quality Testing Requirements

Emissions

Product / Direct Emissions
Life Cycle / Indirect Emiissions
Carbon Dioxide and Global Warming Potential:
Embodied Carbon
Accounting and Emission Reporting

Carbon Offsets
Acidification
Nutrient Pollution and Eutrophication
Photochemical Ozone Creation (Smog)

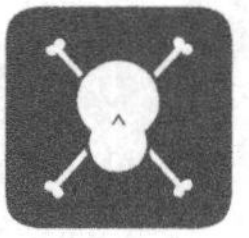

Toxicity and Human Health

Hazard versus Risk
Exposure Pathways
Levels of Toxic Impacts
Effects of Toxins
Exposure Limits

Evaluation and Control:
Bans on Specific Chemicals
Chemical Red Lists
Six Chemical Classes
Characterizing, Optimizing, and Managing

Social Accountability

Labor and Human Rights:
Child and Forced Labor
Discrimination
Working Hours
Unions and Collective Bargaining
Management Processes

Social Impact Indicators
Safe Working Environments
Requirements for Grievance Mechanisms
Third Party Audit
Animal Welfare

Fig. 3.1 Sustainable impacts and environmental initiatives organized into six categories

3.2 Life Cycle Thinking

Life cycle thinking divides a product's life into stages. For each stage different environmental and social impacts can be evaluated. Then different strategies are applied for each stage to minimize negative environmental and social impacts. The most important stages are:

- Extraction of raw materials
- Manufacturing
- Use
- End of life, including recycling

At the beginning of each chapter describing the six sustainability categories, there is a diagram showing which life cycle stages are most affected by the specific sustainability category. While the number of phases in a product's life cycle varies depending on the complexity of the product, there are several general phases (Table 3.1). When the entire product life cycle is considered, this is referred to as "cradle to grave" or "cradle to cradle," cradle meaning the place of birth and grave

Table 3.1 A product's life cycle can be broken into four broad categories: extraction, manufacturing, use, and end of life

Phase	Description	Example(s)
Extraction "Cradle"	The phase during which raw materials are extracted as new or recycled ingredients, including feedstocks	Mining and quarrying Oil drilling Timber harvesting
Manufacturing "Gate"	A phase made up of a number of steps, including material processing, manufacturing, assembly, and fabrication. During this phase a product is transformed from ingredients into a usable building material or product ready for purchase by a consumer	Processing oil into plastic pellets and eventually carpet fiber Processing timber into lumber and paper Refining bauxite into aluminum oxide, smelting aluminum oxide
Use	The duration that a material or product is part of a building. This includes the lifespan of appliances and equipment. This stage also includes maintenance and replacement	The period of time that interior and exterior materials are installed in a building The furniture, appliances, and systems that service a building's users
End of Life "Grave"	When a material or product is removed or decommissioned and taken to a landfill or recycled back into ingredients	Furniture taken to a landfill Recycling plastic bottles into pellets that can be processed into synthetic fibers

being a product's final resting place. Often, the portion of the product's life cycle from extraction to the end of the manufacturing stage is isolated for measurement and analysis – this is referred to as "cradle to gate."

Measuring a product's impact across a life cycle is difficult. Inputs include ingredients, energy, and water. Outputs include waste and emissions. Inputs and outputs exist at every step of the life cycle (Fig. 3.2). Additionally, social and health impacts for workers and consumers run parallel to these environmental impacts at each step of the life cycle.

3.3 Life Cycle Assessment

Life cycle thinking divides a product's life into stages, describing the general approach of considering the different stages of a product's life holistically when determining its sustainability. The specific method for calculating environmental impacts across these stages is called life cycle assessment (LCA). LCA is standardized by the International Organization for Standardization (ISO). ISO provides guidance on LCA through several international standards:

- ISO 14040: Environmental Management – Life Cycle Assessment – Principles and Framework
- ISO 14044: Environmental Management – Life Cycle Assessment – Requirements and Guidelines

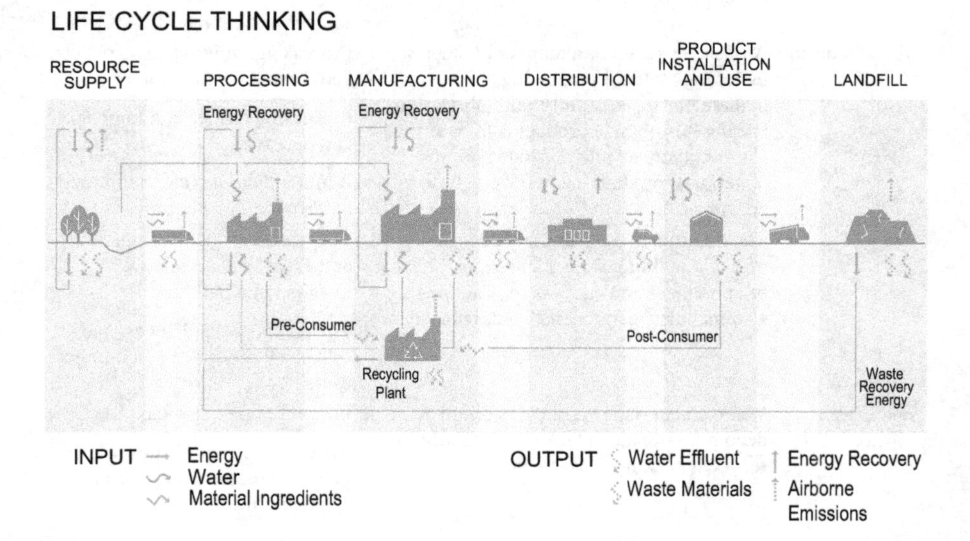

Fig. 3.2 Life cycle thinking

LCAs calculate the inputs (materials, energy, water) and outputs (waste, emissions) throughout a product's life cycle and convert them into their impact on climate change and other indicators of damage to ecosystems and human health. Across life cycle stages, life cycle assessment evaluates:

- Material resource use and composition
- Potential emissions to air and water
- Quantity and character of energy use, including for transportation
- Waste disposal

Life cycle assessments provide comprehensive and quantitative impact estimates. They simplify complex processes and allow for the identification of the largest opportunities for improvement across a product's life cycle. This identification can help manufacturers, regulators, and other decision-makers to focus financial resources and time on decreasing environmental impacts, in particular decreasing emissions from the highest impact ingredients or processes. However, LCAs are not a complete tool for the assessment of a product. LCAs do not excel at measuring human health impacts and social accountability related to a product.

Life cycle assessment initially appeared as an exercise by companies to compare different methods of packaging in the 1970s [1]. The comparison looked at both the materials being used and their environmental impacts at the beginning of the environmental movement. During the next two decades, the focus on LCA lost favor in the United States but gained uptake in Europe.

Today, a manufacturer or company typically hires an internal or third-party evaluator, known as a LCA practitioner, to collect data and complete a LCA for a product. Depending on the goal and scope of the LCA, the result may be an internal study used to inform product design and areas for process improvements, or it may be an externally published document used to communicate the impact of a product to consumers and governmental agencies. Completion of a LCA is generally voluntary, and disclosure of life cycle assessment data is typically not regulated. LCAs are now encouraged or required by a number of green building certification programs, either through the publication of an Environmental Product Declaration (EPD) or as a standalone LCA (EPDs are described in Chap. 11).

3.3.1 Life Cycle Stages and System Boundaries

ISO 14044 defines four key stages in a life cycle: (A) product, (B) use, (C) end of life, and (D) reuse, recovery, and recycling (Table 3.2). Within each of these stages are more refined steps. For clarification, A1: Raw material supply is analogous to the extraction phase described in Table 3.1.

Establishing the system boundary of a LCA defines what stages are included in evaluation and calculation. A "cradle to gate" LCA sets system boundaries from A1 to A3, whereas a "cradle to grave" LCA sets its boundaries at A1–C4. Defining what is excluded from the system boundaries is as important as defining what will be

Table 3.2 ISO 14044 life cycle stages

Stage A	Stage B	Stage C	Stage D
Product Stage	Use Stage	End of Life Stage	Reuse, Recovery, and Recycling
A1: Raw material supply *A2:* Transport between extraction site and manufacturing *A3:* Manufacturing *A4:* Transport between manufacturing site and project site *A5:* Construction and installation process	*B1:* Use *B2:* Maintenance *B3:* Repair *B4:* Replacement *B5:* Refurbishment *B6:* Operational energy use *B7:* Operational water use	*C1:* Deconstruction and demolition *C2:* Transport *C3:* Waste processing *C4:* Disposal	

included: many studies exclude stages such as A5 or B7 because data is not available. The system boundary also describes to what level of detail the product's impact is assessed, or how "far back" the LCA goes. Here are some examples of information that is typically excluded from a LCA based on its system boundary definition:

- Manufacture of the vehicles and equipment used in stages A and C
- Energy and water use related to employees (i.e., factory workers, construction workers, etc.) for commuting and using common facilities such as bathrooms and eating areas

Excluding data from a LCA is acceptable if the LCA assessor identifies the gaps or missing data in the published LCA and does not compare products with differing system boundaries.

3.3.2 The Four Steps of Life Cycle Assessment

ISO 14044 identifies four key steps in a life cycle assessment:

- Step 1: Goal and scope definition
- Step 2: Life cycle inventory analysis
- Step 3: Life cycle impact assessment
- Step 4: Interpretation

Life cycle assessment is an iterative process. This means that after completing Step 4, the assessment team may need to return to Steps 2 and 3 or completely start over and redefine the goal and definition of the LCA in Step 1. Each step is often repeated multiple times before a life cycle assessment is complete. For example, it may become apparent after looking at the initial results that additional life cycle inventory data is required, or that the LCA requires a different unit of measure (also known as a functional unit) to allow comparison.

3.3.2.1 Step 1: Goal and Scope Definition

This step specifies the goal and scope of the study. During this step, the organization and assessor performing the LCA must define:

- Intended application and audience
- Whether the study will be used to compare the product to other products
- Which components will be assessed
- Which life cycle impact categories will be assessed in Step 3
- System boundary or stages that will be assessed
- Functional unit of the product

The functional unit defined in Step 1 of a LCA enables comparison between products. The functional unit addresses the following characteristics:

- Quantity: The number and type of units being assessed, such as 1 square foot of flooring or 1 gallon of paint
- Function: The intended use of a product, such as for flooring, structural use, protection from weather, etc.
- Performance: Durability, resistance to fire or other elements, maintenance requirements, and other characteristics
- Service Life: The expected length of the use stage of a product before it is disposed of or recycled

Defining the quantity, function, performance, and service life of the product being assessed determines how a product LCA can be used for comparison with other products in the future. For example, if the LCA of Products A and B demonstrates that they have the same total environmental impact, however, Product A has a service life of 5 years, and Product B has a service life of 10 years, through calculation Product A has twice the environmental footprint as Product B, because it would have to be replaced once to last for the same reference period.

This issue is similar for other aspects of the functional unit. Consumers cannot compare two product LCAs if quantities differ or have different functions. For example, comparing the environmental impact of 1 square foot of carpet tile to 1 square foot of ceiling tile would not be useful. If the ceiling tile has a lower impact than the flooring, that does not mean it can be used as flooring. By accurately defining the functional unit of a LCA in Step 1, future consumers can determine whether they may use the LCA for comparison.

3.3.2.2 Step 2: Life Cycle Inventory (LCI)

The second step of a life cycle assessment (LCA) is the collection and quantification of inputs and outputs for a given system throughout its life cycle to create the life cycle inventory. Inputs typically include the quantity and type of raw materials, the amount of water and energy used, and the transportation distance and method. Outputs are also collected for each process, including the quantity and type of waste generated during each stage. Each item in the life cycle inventory must include the unit and phase. For example, one inventory item could be 10 kilograms (kg) of steel used in Stage A3: Manufacturing or 100 megajoules (MJ) of electricity from the Southwestern United States' electrical grid used in Stage A1: Raw material supply.

Collecting data for a LCA is time-consuming and expensive at best. At worst it may not be possible if a manufacturer does not have sufficient information regarding their supply chain. As a result, generic data is often used in lieu of specific data. This can limit the value and specificity of the assessment. Where data is not available, the LCA practitioner must decide whether they should exclude the stage or component from the system boundary of the LCA or if they should use data collected from a similar process or product to fill the data gap.

LCA practitioners also have to judge the quality and completeness of data provided to them, as they rarely collect the data firsthand. Common issues related to data quality include:

- Allocation: Manufacturers often produce more than one product in a facility or may produce pieces of a product across several facilities. In each case, to complete LCA the portion of the total resources and environmental impacts must be broken down and allocated or associated with each product, for instance, how much of the facility's energy use is dedicated to a specific product. Incorrect allocation leads to poor data quality.
- Age: Outdated data may not account for newer technologies or processes.
- Geographic Relevance: Data representing a manufacturing process for one country may not be representative of the same process in a different country due to different manufacturing techniques, electrical grids, resource availability, and other differences.

Making assumptions and using imperfect data is a necessary part of performing LCAs. The LCA practitioner must make educated decisions about when it is or is not acceptable to use data. They may look at similar data points to see if the order of magnitude of the data provided for a process or material is within the range. In Step 4, each assumption may also be tested using sensitivity analysis to determine if more accurate data needs to be collected to have meaningful results.

3.3.2.3 Step 3: Life Cycle Impact Assessment (LCIA)

During the third step of a LCA, the impact of each input and output associated with a product is summed and expressed as life cycle impacts. Specifically, the LCA practitioner must identify which emissions should be included in the calculation of each impact and characterize them accordingly.

Equivalent units are used to express life cycle impacts. This means that a specific impact is expressed relative to the impact of one unit of a specific substance. For example, the impact of a wide variety of greenhouse gases is expressed by their impact relative to 1 kilogram (kg) of carbon dioxide (CO_2). This is written as 1 kg CO_2eq or 1kgCO_2e. The process of characterizing emissions with equivalent units helps translate hundreds of unique emissions into just a few impacts that can be compared and tracked more effectively (Fig. 3.3).

Life cycle impact assessment (LCIA) is completed by applying factors to the inventory analysis from Step 2 to calculate the environmental impact of each input and output. This is a simplified version of the LCIA process that occurs during Step 3:

Inventory item	x	Impact factor	=	Total [equivalent] impact
kg of steel	x	*$kgCO_2$ per kg of steel*	=	*$kgCO_2eq$*

The impact factor applied to each inventory item depends on the characterization method selected for the LCA. A characterization method is a framework published by a company or governmental agency that is used to define life cycle impacts and assign numerical factors to calculate each impact. Characterization methods are established by governmental or non-governmental organizations to help practitioners consistently relate emissions and impacts. In 2002, the US Environmental Protection Agency (EPA) introduced the Tool for the Reduction and Assessment of Chemical and Other Environmental Impacts (TRACI), a characterization method in the form of a database that evaluates chemicals for possible environmental threats [2].

The most commonly tracked environmental impact categories for building materials and buildings are acidification, eutrophication, global warming potential, smog formation potential, and depletion of non-renewable resources. Some impact categories, such as human health and ecotoxicity, are less commonly included in a LCA

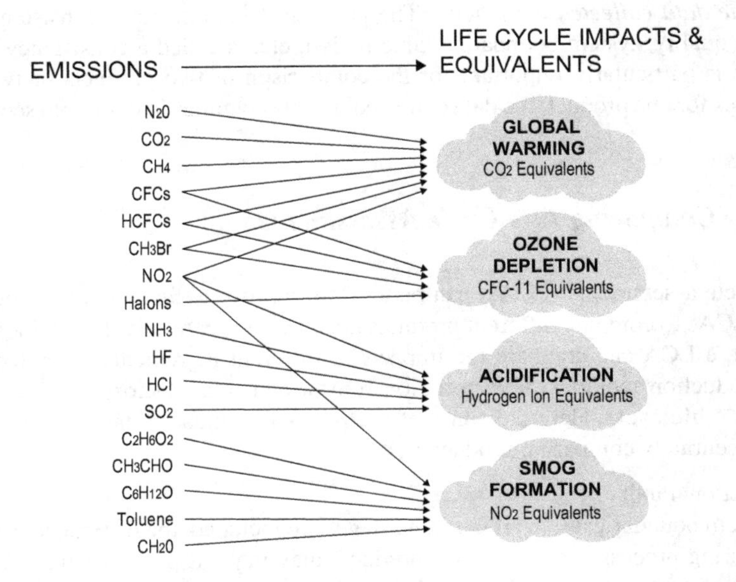

Fig. 3.3 Examples of how life cycle emissions (left) are converted to equivalent units to understand common life cycle impacts

because the methods for calculating and characterizing impacts are less standardized [3]. To understand more about each of these environmental impacts, refer to Chaps. 4 through 9.

3.3.2.4 Step 4: Interpretation

In this final phase, the results of the LCA are checked and evaluated to determine whether they are reliable and ready for publication. Here are some of the questions that may be asked during this step to determine the reliability of the study:

- *Are assumptions or poor data quality skewing the results?* Assumptions are a necessary part of every LCA. The assessor performing the LCA may recalculate the analysis with a range of numbers in place of one estimate to see if it has a large impact on the data. If the results show that the particular inventory item is a large percentage of the overall impact, then the assessor will collect more data to apply a more accurate estimate. For example, the exact distance traveled for Stage A2 is unknown, but the assessor estimates 300 miles. During Step 4, they might recalculate the analysis with 250 miles and 350 miles to see if this changes the relative percentage. This process of testing the impact of more unreliable data on the final results is called a sensitivity analysis.
- *Are there important gaps in the data included in the study?* Does the inventory analysis in Step 2 include all relevant items for the stages set in Step 1? Looking for missing or incomplete areas for data collection to meet the goal and scope of the study is referred to as a gap analysis or completeness check.
- *Is the data collected consistent?* The process of looking for inconsistencies in data quality, system boundaries, time period, etc. is called a consistency check. This is particularly important for the comparison of two products or two processes for one product: the data collected must be symmetrical for both scenarios.

3.3.3 Comparing Life Cycle Assessments

Life cycle assessments (LCAs) can be used to improve an individual product, but using LCAs to compare different products must be done carefully. For an individual product, a LCA can compare the impacts of different ingredients, manufacturing and production methods, transportation routes, and other factors specific to one product's life cycle. Here are some of the issues that must be taken into account when accurately comparing products:

- Functional units.
- System boundaries: Even if the same stages are included, the difference in manufacturing processes between two products may make it unclear what life cycle stages or processes must be included to be comparable. To compare a cork floor

to carpet tile, or a stone countertop to one made of laminate, is more technically challenging than the comparison of two carpet tiles.

- Data quality and assumptions.
- Impact Evaluation: The scope of how much damage actually occurs varies widely depending upon the condition of the environment or health of a population into which this material is released.

References

1. Bjørn, A., Owsianiak, M., Molin, C., & Hauschild, M. Z. (2018). LCA history. In: M. Hauschild, R. Rosenbaum, & S. Olsen (Eds.), *Life Cycle Assessment*. Cham: Springer. https://doi.org/10.1007/978-3-319-56475-3_3. Accessed 9 Sept 2020.
2. US Environmental Protection Agency (EPA). (2016). *Tool for reduction and assessment of chemicals and other environmental impacts (TRACI)*. https://www.epa.gov/chemical-research/tool-reduction-and-assessment-chemicals-and-other-environmental-impacts-traci. Accessed 8 Sept 2020.
3. Simonen, K., Huang, M., Rodriguez, B., & Todaro, L. (2019). *Life cycle of buildings: A practice guide*. The Carbon Leadership Forum, Department of Architecture, University of Washington. https://carbonleadershipforum.org/projects/lca-practice-guide/. Accessed 1 Sept 2020.

Chapter 4
Resource Use

4.1 Introduction

Resource use refers to the ingredients that are used in a product, such as wood, metal, plastic, and other materials. In some cases, using energy and water may also be referred to as resource use, but this chapter focuses on strategies related to a product's raw materials and ingredients. The considerations for resource use are centered around several initiatives and issues (Fig. 4.1). Questions include:

- How can a manufacturer source more sustainable ingredients?
- Can the product be redesigned to use less materials or better ingredients?
- What happens to the product at the end of its use? Can it become a resource for new products?

Within a product's life cycle, resource use strategies primarily fall into three life cycle stages: extraction, manufacturing, and end of life. To a lesser extent, resource use considerations and strategies are present throughout the product life cycle (Fig. 4.2).

4.2 Selecting Sustainable Ingredients

Manufacturers must decide which ingredients go into a product. These ingredients can lead to environmental and social impacts that are highly interconnected including impacts across energy, water, toxicity, emissions, and social accountability. All of these impacts are described in the next five chapters. For example, when a manufacturer selects a specific type of plastic, that decision determines whether the product will be recyclable and how much energy and water will be required for processing. Choosing between wood and metal for a chair determines the types of

H. R. Roth et al., *The Green Building Materials Manual*,
https://doi.org/10.1007/978-3-030-64888-6_4

INITIATIVES AND IMPACTS: RESOURCE USE

Resource Use

Non-Renewable Resources:
Ecosystem Impacts
Local Sourcing
Quarrying, Mines, & Wells
Conflict Resources
Renewable Resources:
Biobased
Rapidly Renewable
Wood Sourcing
Carbon Sequestering

Circular Feedstocks:
Reused/Reclaimed Content
Recycled Content
Dematerialization
Waste Recovery and Circularity:
Biodegradability/Compostibility
Recyclability
Designed for Disassembly
Waste Recovery/Take Back Program
Reduction in Packaging

Fig. 4.1 Initiatives and impacts for resource use

RESOURCE USE | WASTE DURING LIFE CYCLE

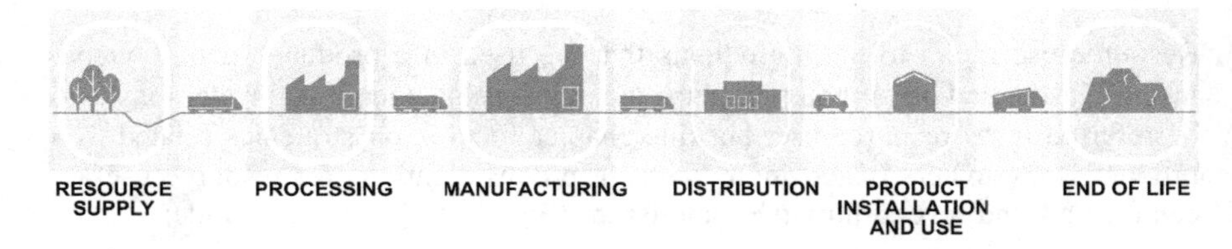

Fig. 4.2 Resource and raw material use throughout the product life cycle

finishes and glues that will be required. Selection of finishes and glues establishes a product's toxicity and human health risk and whether it generates product emissions.

Additionally, choosing a material that must be extracted has different consequences compared to choosing a material that is recycled. Different considerations must be made when choosing a material that is non-renewable or taken directly from the earth, compared to using a material that is grown and renewable (Table 4.1).

4.2.1 Resource (Raw Material) Extraction Sites

When using a new material, the first impacts occur at its extraction site, when the material is either extracted from the earth as a non-renewable resource or grown or harvested as a renewable resource.

4.2.1.1 Ecosystem and Biodiversity Impacts

Raw material extraction sites:

Table 4.1 Resource (raw) material sourcing issues

Key Sourcing Issues	Description	Specific Area of Focus
Raw Material Extraction Site *Non-renewable Resources* *Renewable Resources*	Where is the ingredient originally taken from? What are the associated sustainable impacts?	Ecosystem and biodiversity impacts Local sourcing opportunities
Non-renewable Resources	These are materials removed from the earth that will not be naturally replenished during a human lifetime For example, stone and sand, metals and minerals, fossil fuel-derived plastics	Quarries' and mines' impacts Conflict resources' impacts
Renewable Resources	These are materials that are grown For example, wood, bamboo, cotton, wool	Biobased content Rapidly renewable content Wood sourcing Carbon sequestering
Reused or Recycled Material *Non-renewable Resources* *Renewable Resources*	When a material is reused, what are the associated sustainable impacts?	Reused content Reclaimed content Recycled content

- Disturb a site's soil.
- Pollute the air and water with emissions.
- Disrupt ecosystems with light and noise.

Manufacturing facilities, construction sites, and other locations also threaten ecosystems. The impacts of an extraction or manufacturing site may be difficult to relate to impacts on a single species. It is often challenging to single out one possible factor or disturbance, especially since many negative effects become apparent over a period of time.

State and federal regulations protect areas with endangered species, restrict air and water pollution, and limit soil movement. These are well suited to preventing broad, distributed ecosystem impact. Raw materials extractions sites that follow these regulations limit impact on ecosystems. However, many raw material extraction sites are not located in the United States and may not follow US regulations.

4.2.1.2 Local Sourcing

The closer a material is sourced to its point of installation, the less environmental impact from transportation the product produces. It is also more likely local residents will benefit economically since a product produced nearby typically provides employment to adjacent communities. However, there is no consistent definition of local sourcing, and the terminology is not regulated by a governmental agency. As a result, individual regulations and standards define the distance for what is considered local, for instance (Fig. 4.3):

- LEED provided credit for materials sourced from a radius of 500 miles to a project's site prior to Version 4.1. The most recent version of LEED (v4.1) has decreased the source radius to 100 miles, and materials produced within a 100 mile radius no longer count directly toward a credit for certification [1].
- The Living Building Challenge requires 20% of building materials source from 500 km, 30% from 1000 km, and another 25% from 5000 km [2].

4.2.2 Non-renewable Resources

Non-renewable resources formed due to geological activity. They formed millions of years ago. To be used, they must be mined, quarried, or otherwise extracted from the earth. Common non-renewable resources include stone, sand, metals, minerals, and fossil fuel-derived plastics. Many of these materials are purchased as commodities and are difficult to track to their original source site. Because of often limited supply chain information, it becomes difficult to know whether these raw materials are responsibly extracted.

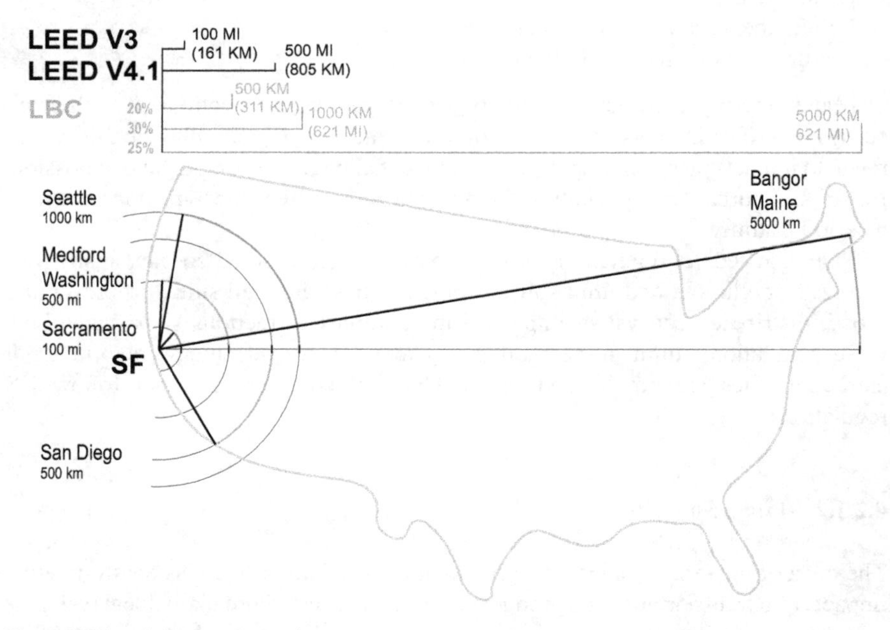

Fig. 4.3 Visualizing local sourcing

4.2.2.1 Quarries, Mines, and Wells

Quarries supply stone, rock, and sand. They are typically an open pit, and nearly all the material that is removed is sold. A mine supplies minerals and metals. Often mines are underground, though some are also open pits. Minerals and metals are extracted as ore and need to be refined. The refining process leaves behind tailings which is the rock that does not have enough minerals or metals to be sold. Tailings contain metals and other toxic substances that can be harmful to the environment once they are brought to the surface of the Earth. Fossil fuels can be extracted using open pit mines, for example, for coal or tar sands. Wells supply oil and water. Quarrying, mining, and wells in the United States are monitored by the Environmental Protection Agency (EPA) and state authorities. These extraction operations must comply with both federal and state regulations.

Quarries' and mines' impacts on ecosystems include all the environmental impact categories that are discussed in Chaps. 5 through 9 (Table 4.2).

Even though non-renewable resources were formed millions of years ago, the extraction process is inherently temporary. A mine or quarry site is established, and materials are removed. When it is no longer economically viable to remove materials, the site is closed. The abandoned operations can have serious impacts on adjacent communities and ecosystems. Reclamation seeks to return a site of operations to a non-toxic and usable condition. This typically requires a capping or sealing process to isolate toxic materials and prevent leakage into ecosystems.

The EPA and local agencies require quarries, mines, and wells to follow regulations in the United States. In addition, several trade groups and organizations have established standards for extraction operations. For quarrying, the Natural Stone Council instituted a certification for sustainably managing quarries with focus on quarrying, processing, and transportation in 2017. For mining several standards for certification are currently being developed including:

- Initiative for Responsible Mining Assurance, Standard for Responsible Mining
- CERA Certification of Raw Materials

Additionally, the International Organization for Standardization (ISO) is developing a global standard for safety and sustainability during operations and during reclamation after a mine closes.

4.2.2.2 Conflict Resources

Valuable minerals, metals, and fuels are distributed in locations across the globe. Unfortunately, some of these resources are located in places of conflict and war. Conflict resources are natural materials that are mined, quarried, or collected to fund war or conflict. In particular, mining operations for coltan, tin, gold, and tungsten are located in war-torn regions. These precious materials are widely employed in semiconductor production, photovoltaics for solar collectors, and other electronics manufacturing [3].

Table 4.2 An overview of environmental impacts related to quarries, mines, and wells

Environmental Impact	Description	Impact
Energy Use	Fossil fuels are mostly used because operations are often remote, and fuel must be brought to the area Mining and quarrying use large amounts of energy	High carbon footprint
Water Use	Mining and quarrying equipment often uses large amounts of freshwater Used polluted water (effluent) is discharged from operations	Water-intensive process Operations may be in water-scarce areas Effluent either is discharged into ecosystems, causing harm, or is held in ponds on site, where it can harm birds and other mammals Holding ponds on occasion fail, allowing effluent to enter the ecosystem and cause further damage
Emissions	Operations create dust and particulates Equipment releases fossil fuel-based emissions Burning excess fuels releases emissions	Carbon dioxide and greenhouse gases Acidification and smog Airborne particulates
Toxicity and Health	Dust and particulates are released into the air Metals and minerals release toxicity into the air, water, and adjacent soils Noise pollution Operations are often 24 hours causing light pollution at night	Toxic to human health Toxic to environment and ecosystems Noise and light pollution disturb ecosystems
Social Accountability	Operations in the United States must follow US labor law Operations outside of the United States may not follow US labor law	In operations outside the United States: *Potential for child labor* *Potential for forced labor* *Worker hazards and safety issues* *Work week hours may be longer than approved by global labor law*

Legislation to limit the purchase of conflict resources has focused upon the Democratic Republic of the Congo. In particular the Dodd-Frank Wall Street Reform and Consumer Protection Act required manufacturers to monitor supply chains and report purchase of conflict resources, though some part of that disclosure was later struck down in court. Currently, the US Department of Energy (DOE) is funding research at universities and national laboratories replacing conflict materials used in photovoltaics and other electronics manufacturing with abundant materials.

4.2.3 *Renewable Resources*

Renewable resources are grown and harvested. They can be either raw materials like wood and fiber or come in the form of a chemical (feedstock) that makes up a plant-based plastic. Renewable materials are either plant-based or animal-based. Plant-based materials inherently have low embodied carbon in comparison to other materials. They draw carbon out of the atmosphere as they grow. However, plant-based materials can result in excess fertilizers and nutrients being released into soil and nearby bodies of water. Additionally, the burning of plant-based materials and any remainders after harvest release their embodied carbon back into the atmosphere. There are several ways to document both the amount of plant-based material in a product and how the agriculture was raised and processed.

In contrast with plant-based materials, animal-based materials have high embodied carbon. Small amounts of an animal product, such as leather, require enormous amounts of grazing land or processed feed. Animal digestion processes release large amounts of methane into the air. Additionally, cleaning and tanning materials such as leather require toxic chemicals that are harmful to human health.

Sustainable initiatives for renewable resources are evaluated through:

- Biobased content
- Rapidly renewable content
- The sourcing methods for wood
- Evaluation for sequestering carbon

4.2.3.1 Biobased Content

Biobased products are commercial or industrial products that are grown or harvested. As a crop, their purpose is not to be used as food or feed. Crops include corn, soybean, and flax. The US Department of Agriculture (USDA) designated the term "biobased" and supports biobased products as alternatives to petroleum-derived products [4]. These materials are usually a chemical component or feedstock, rather than filler, fiber, or aggregate. Examples include:

- Biobased plastics
- Biobased adhesives
- Biobased paints and sealants
- Packaging

The 2002 USDA Farm Security and Rural Investment Act, authorized by President George W. Bush (reauthorized in 2014), established law allowing the USDA to regulate biobased products. The law was established to support agriculture. The goal was to create new opportunities for farm commodities, replacing petroleum-based products with agriculturally sourced products. The USDA monitors biobased content across 135 categories of products. Federal agencies must show a preference for purchasing biobased in their product purchasing programs.

Products may be certified as biobased or report a percentage of biobased content through packaging or product literature. Content is verified by third-party testing to ASTM D6866-18/ISO 16620-2 using radiocarbon analysis. Sustainable building certifications such as Leadership in Energy and Environmental Design (LEED) and the Living Building Challenge (LBC) have additional requirements beyond the USDA's regulations to ensure that only biobased products with reduced environmental impacts are used on projects. For instance, LEED v4.1 requires that materials are tested using ASTM Test Method D6866, be legally harvested, and meet the Sustainable Agriculture Network's Sustainable Agriculture Standard [5].

4.2.3.2 Rapidly Renewable Content

Rapidly renewable is a term first defined by the US Green Building Council (USGBC) for LEED certification. The term defines a biological material that can regrow for harvest in 10 years or less. While biobased materials are processed down to the chemical level, rapidly renewable materials are used at a higher, less processed, value – such as finish, fiber, filler, or aggregate. Examples include corn, bamboo, wool, and cork.

Unlike biobased materials, rapidly renewable content is not federally legislated. Third-party organizations may verify content if a manufacturer is seeking certification for rapidly renewable content. In the nearly 20 years since LEED began, LEED and many other certifications have changed their requirements, intensifying their expectations on sustainable initiatives. In the latest version of LEED's building certification, LEED v4.1, rapidly renewable content is no longer valued toward earning credits for certification. This is true of other green building certifications as well.

4.2.3.3 Wood Sourcing

Forest management on federal lands first began in 1876 when Congress authorized the USDA to evaluate the conditions of the forests across the United States. Today, several organizations certify forest management on private and public lands within the United States and internationally. These certifications focus on how wood as an agricultural product is managed. The certifying organizations:

- Establish criteria for forest management, such as:
  - Maintenance of biodiversity and protection of wildlife habitat
  - Soil and water protection, including erosion, during the planting, growth, and harvest of managed forests
  - Establishing schedules for replanting, reforestation, and harvest
- Work with third-party auditing organizations to confirm compliance with standards

- Provide documentation on compliance and supply chain from processing and manufacturing to ensure documentation for chain of custody

Sustainable Forestry Certifications

There are two primary sustainable forestry certifiers: the Forest Stewardship Council (FSC) and the Sustainable Forestry Initiative (SFI). Both organizations seek to protect forests from environmental impacts when trees are harvested. In both cases, the organizations publish proprietary standards, provide a certification that adheres to their standard, and affix an ecolabel of the organization's name to products that have earned the certification. While these two organizations represent themselves similarly on websites and in organization literature, they use different criteria by which they manage forests [6].

The Forest Stewardship Council (FSC) is a non-profit membership-led organization, with more than 380 million acres certified. The FSC certification originated from a group of non-governmental organizations' (NGOs) concern about tropical deforestation in the 1990s; these NGOs worked with industry leaders to establish a voluntary certification system. In 1995, the Sustainable Forestry Initiative (SFI) was launched as a voluntary conduct code for members of the American Forest and Paper Association.

Because of different forestry management practices, FSC has historically been preferred for wood certification by both federal agencies and LEED. Recently SFI has directly lobbied to receive equal consideration. The back and forth reveals the stakes involved in receiving endorsement by federal agencies and independent standards.

In 2015, the DOE and the EPA issued an interim recommendation for federal procurement programs that allowed FSC lumber and excluded SFI lumber [7]. In LEED Version 3, USGBC mandated that a minimum of 50% of wood-based materials be FSC certified to earn the LEED MRc7 credit. LEED was silent on SFI wood. In 2016 the USGBC issued a pilot alternative compliance path (ACP) to prevent illegal forest products in LEED buildings [8]. The pilot program recognizes both FSC and SFI and will become a required prerequisite in the future. In LEED v4.1, wood sourced from FSC or a USGBC approved equal can help to fulfill the Building Product Disclosure and Optimization – Sourcing of Raw Materials, opening the door to SFI approval.

In June 2016 a bi-partisan group of senators wrote to the DOE asking to reverse course and include SFI and the associated American Tree Farm System as suppliers of products for federal purchasing programs [9]. The group specifically cited the USGBC's decision to include SFI in its new pilot compliance path. No action has yet been taken. Federal procurement programs control millions of dollars in purchasing.

4.2.3.4 Carbon Sequestering (Carbon-Storing) Materials

Carbon sequestering materials and carbon-storing materials are materials that capture carbon dioxide during the growth phase of their life. Carbon is stored during their product life, preventing it from being released into the atmosphere. There are also engineered materials that sequester carbon over the life of the material. This includes formulations for certain types of cement or for oxidized metal coatings.

Research on biobased carbon-storing materials is relatively new, and there is concern that carbon sequestration claims may be green washing if not backed by third-party certified sustainable harvesting practices. Engineered materials that sequester carbon over their product life also pose an additional dilemma. Engineered carbon sequestering materials are often developed using new engineering or manufacturing processes. It is difficult to know whether these products will have negative unintended consequences in the future.

4.2.4 Circular Feedstocks: Reused, Reclaimed, and Recycled Resources

Since the Resource Conservation and Recovery Act of 1976 (RCRA), federal law has encouraged manufacturers to replace virgin or new materials with reused resources. These materials contain:

- Reused material content
- Reclaimed material content
- Recycled material content

4.2.4.1 Reused Material Content

Reuse of materials is an age-old practice. From ancient times through the middle ages, spolia, or elements from demolished buildings, were used in new construction adding value by providing ornament or embellishment.

- *Reused material content* means a material or product is used an additional time for its original purpose.
- *Repurposing* a material retrieves a material from a previous use or installation and then reuses that material for a different purpose or use. Repurposing a material may mean that it is resized, refinished, or connected to other elements in a different or unexpected way.
- *Upcycling* reuses a material but adds some value through printing on it, refinishing it, or adding additional material or components, so that the reused material is now seen as worth more.
- *Refurbishing* a material or product rehabilitates it before reusing it.

Constructing the Model A and Model T: Renewable Resources and Repurposing

Spanish moss is an epiphyte. It does not need soil and gets its nutrients from moisture in the air. It attaches to the bark of tree limbs. It grows well in the cypress that fills the swamps of southern Louisiana. For years Cajuns harvested Spanish moss out of the Atchafalaya Swamp. They processed it to become batting for mattresses, pillows, and quilts. Henry Ford learned about Spanish moss and ordered it as fill for his Model T car cushions.

He required that the Spanish moss be shipped in very specifically sized cases made out of cypress. Unbeknownst to the Cajuns, he then used that correctly dimensioned cypress packaging for the running boards and dashboards of the first Model A and Model T cars.

Henry Ford also used timber from other parts of the country. Always interested in making use of resources to their fullest, Henry Ford looked for a use for leftovers from fabricating the wood into running boards and dashboards. A family member by marriage was named Kingsford. Henry Ford worked with Kingsford to develop a side business, processing the wood into charcoal briquets. Charcoal briquets are still sold under the Kingsford name [10].

4.2.4.2 Reclaimed Material Content

Reclaimed materials, also referred to as salvaged materials, are reused materials that were previously installed in a building or project [11]. Upon deconstruction or dismantling, materials are salvaged for use in another project. Alteration may include re-sizing, refinishing, or adaptation. If a reclaimed material is used in a different way, it is referred to as repurposed. Common reclaimed materials include structure and cladding materials such as bricks, slate, tile, and timber as well as elements, such as doors, window frames, and fireplaces.

Documentation establishing chain of custody is important in maintaining an understanding of the history of the material and confirming that it was salvaged legally. Chain of custody also allows for investigation into health risks, such as contact with lead paint or other toxic substances that were not previously regulated.

4.2.4.3 Recycled Content

In 1955, *Life* magazine published an article on “Throwaway Living” emphasizing how single-use items decreased household chores. The article listed items such as disposable skillets, draperies, and hot pads. It even suggested disposable diapers were responsible for an uptick in birth rates [12]. Advertising campaigns, product literature, and popular culture increasingly declared that single-use items were an expression of modern living. Ten years after the publication of "Throwaway Living",

Table 4.3 Types of recycled content

Type of Recycled Content	Description
Post-consumer Content	Materials diverted from the waste stream *after consumer use*
Pre-consumer Content or Post-industrial Content	Materials diverted from the waste stream *during the manufacturing process* Pre-consumer materials must be actually destined for waste. They cannot be extra raw materials, such as materials trimmed off a product, which would be normally collected and reincorporated into feedstock

Gary Anderson won an art contest to increase awareness for the environment with his now internationally recognized Reduce, Reuse, Recycle symbol, in the form of a Mobius loop. In 1981, Woodbury, New Jersey, became the first city to require curbside recycling. As of 2020, 59% of the US population has recycling provided as part of their trash service; unfortunately, only about one third of the recyclable materials generated by households is actually captured by curbside recycling [13].

Recycled content is defined as material diverted from the solid waste stream, reconditioned, and used as material feedstock. Recycled content often requires processing down to a chemical or building block level before being reused as feedstock. Certifications distinguish between two types of recycled content: post-consumer content and pre-consumer or post-industrial content (Table 4.3).

The federal government began regulating waste removal in the 1960s. In 1965, Congress enacted the Solid Waste Disposal Act (SWDA), which provided a structure for states to control waste and set minimum safety requirements for landfills. The 1976 Resource Conservation and Recovery Act (RCRA) amended the SWDA. This law called for research on the economic viability of reused materials, development of guidelines for government purchasing of materials with reused content, and identification of markets for using reused materials. The EPA led the RCRA's initiatives.

Resource Conservation and Recovery Act

The 1976 Resource Conservation and Recovery Act replaced and amended the SWDA [14] (Table 4.4). The law introduced a framework for conserving materials, linking conservation with economic development. It set the goal "to separate usable materials from solid waste" in order to "reduce dependence of the United States on foreign resources and reduce the deficit in its balance of payments" [14]. It is interesting that this first framework for conservation had not yet adopted the term "recycled."

The act provided directives to do the following:

- Set standards for recovered materials specifications.
- Find markets for recovered materials.
- Assess the economic opportunities of recovery facilities.
- Make changes to government procurement standards to allow materials with reused or recycled components.

Table 4.4 Directives of the RCRA

Directive	Description
Set standards of quality for recovered materials	Develop a classification framework
	Set standards for physical properties
	Set standards for chemical properties
	List virgin materials that can be replaced by recovered materials
	Provide applications to industrial, commercial, and governmental uses
Find markets for recovered materials	Geographic location of existing or potential markets
	Economic and technical obstacles to the use of recovered materials
	Encourage new uses for recovered materials
Assess economic viability of recovery facilities (later called recycling centers)	Assess the profitability of resource recovery facilities
	Develop a database of existing facilities with companion analysis for future investors
Change procurement requirements to allow recovered materials	Remove language from procurement documents that exclude recovered materials
	No longer require that goods be manufactured from virgin materials
	Require recovered and reclaimed materials be used as much as possible without threatening the usability of the product

In 1992, the Federal Trade Commission (FTC) published definitions for recycled content in its Green Guides. Today, the EPA (regulating the RCRA) sets recycling goals, and the Department of Commerce (DOC) finds markets for recycled goods. Additionally, local city and state governments establish legislation and regulation for landfill bans and recycling goals. Numerous volunteer-based programs also exist. The LEED rating system provides points toward certification for recycled content in building materials and for decreasing or recycling construction waste [15].

Circular Economy

With emphasis on the circular economy, the recycling process is being reconsidered. It is no longer good enough to simply recycle something, without knowing what the path of recycling is. It is now important to trace how that recycled good is used. The circular economy seeks to eliminate all waste and to return as much used material as possible to feedstock. Additionally, emphasis is placed on retaining higher, less processed, value – such as fiber, filler, or aggregate – rather than reprocessing down to the chemical and molecular level.

San Francisco and Zero Waste

Across the country, the EPA calculates that municipalities average 35% diversion from landfill through recycling and composting programs [16]. In 2003, San Francisco led the nation by setting a goal to divert 100% of its waste from landfills by 2020.

In 2001, San Francisco had already achieved the goal of diverting 50% of its waste, which was a percentage mandated by a statewide California law passed in 1989 [17]. The 2003 resolution and accompanying 2020 deadline sought to reduce household waste to zero [18].

By 2018, San Francisco had reduced household waste by an additional 50%. This was achieved through an extensive sorting program and a three-container system for waste: trash, recyclables, and materials destined for compost, including food scraps. However, the city could not attain the zero-waste goal by 2020. Mayor Breed established a new goal in 2018 to reduce all waste by 15% and reduce landfill waste by 50% [19]. The difficulty was that San Francisco's recycling and composting programs had become so efficient that now consumer and manufacturer behavior needed to be targeted. The city is currently developing different strategies for different groups (Table 4.5).

In addition to targeting household waste, San Francisco also regulates construction and demolition debris [20]. Many cities have ordinances on diverting construction waste from landfills. San Francisco's regulation is considered stringent. Adopted in 2006, the regulation states that all materials that can be separated must be taken to reuse/recycling facilities. For mixed construction and demolition waste, a minimum of 65% of debris must be diverted from landfill. An example of mixed debris is metal insulating panels. Insulating foam is attached to the metal with adhesive. These types of materials cannot be easily disassembled and sorted.

Table 4.5 San Francisco zero-waste strategies

Group	Strategy
Households	Trash bins have been reduced in size by 50% and are now half the size of recycling and composting bins
	Contaminated trash, recycling, or compost results in fines or suspended discounts
Apartments, Hotels Commercial Offices Hospitals, Universities	A new law requires trash 75% uncontaminated, recycling 90% uncontaminated, compost 95% uncontaminated
	The city audits waste
	Those groups that fail audits will be required to hire waste sorters
Proposal for Plastic Manufacturers	Tax 1 cent per plastic package
	Ban Styrofoam food packaging
	By 2030 all packaging be recyclable, reusable, or compostable

4.3 Designing Products to Minimize Resource Use: Dematerialization

The most straightforward strategy for reducing the environmental impact of any material is to use less. When products are designed to require less materials while maintaining (or enhancing) performance, this is referred to as dematerialization. Even a small reduction can have a huge impact, particularly when multiplied across the sale of millions of units. Optimization may entail:

- Decreasing a product's mass
- Using fewer materials by percentage
- Decreasing the complexity of a product, reducing components

Dematerialization is often a product of technological advancement. When new technologies are initially invented, their design may be less efficient, both in energy use and resource use. After introduction, new iterations of design increase efficiency. For example, the mass of light-emitting diode (LED) lamps decreased 50–60% from 2012 to 2015. This significantly reduced the environmental impact of a single lamp without changing its function or efficiency. LED lamps also increased in efficiency during the same time period [21].

In other cases, manufacturer innovation is required to dematerialize a product. For example, the sheetrock manufacturer USG designed lightweight sheetrock panels that maintain their function while significantly reducing their water, energy, and resource use. Dematerialization can have co-benefits outside of the environment: these panels are also easier and safer for on-site installation.

If something that is a product is re-engineered as a service, then the amount or number of materials needed for the product is reduced to zero. At the scale of construction, an example would be a bank decreasing its number of bank branches and instead offering services only through the Internet.

4.4 End of Life: Waste Recovery and Circularity

When a product is discarded, steps can be taken to decrease waste and maximize the use of the remaining materials (Table 4.6). The EPA waste hierarchy pyramid along with the European waste framework directive of 2008 provides a structure for decision-making from greatest impact to least [22]:

- Source reduction and reuse
- Recycling and composting
- Energy recovery
- Treatment/disposal

Table 4.6 Waste recovery strategies

Strategy	Description
Source Reduction and Reuse	Reuse a product instead of putting it in the waste stream
	Design a product to minimize the amount of waste it creates at its end of life
	Minimize the excess material a manufacturer needs to make a product
Recycling and Composting	Recycle a material for reuse, whether as components or returning it into material feedstock
	Compost a material to create organic nutrients
Energy Recovery	Use a method such as burning to recover energy from a product
Treatment/ Disposal	Shred or compact a material to make it smaller in size, taking up less space in a landfill, or more easily biodegradable
	Develop municipal waste programs such as "Pay as You Throw" which, instead of charging a steady cost for waste management, charge based on volume or weight, providing economic incentive to decrease waste

The strategy for waste management for one product will not be the same for another. A manufacturer or waste management service must evaluate what is the best solution for a given item, or family of items.

4.4.1 Circularity

The term "product circularity" defines and documents the process of returning waste and used materials into the supply stream to become new products. The goal of the circular economy is to eliminate all waste. The circular economy defines everything as "a resource from something else." This idea was first described in the seminal book *Cradle to Cradle*, written by German Chemist Dr. Michael Braungart and US Architect William McDonough in 2002 [23]. Since 2002, the concept has become widely used and embraced by the building industry. Organizations like the Ellen MacArthur Foundation seek to provide resources for implementing the circular economy. The foundation also works with businesses, government agencies, and institutions to establish policy and public-private partnerships to encourage product circularity [24]. Companies increasingly are filling specific gaps or missing infrastructure within the circular economy. For example, TerraCycle has established programs to reuse, upcycle, and recycle difficult to recycle waste streams, from textiles and candy wrappers to clay gardening pots [25].

Manufacturers use the terms product circularity and circular economy in product literature and statements on corporate policy.

Built Positive

Recently, industry leader Cradle to Cradle (C2C) launched the Built Positive movement [26]. This effort focuses on the return of all materials at the end of life back into the supply chain.

To accomplish this, C2C emphasizes:

- "Circular Design": Everything is feedstock for some other product, and all waste should be eliminated.
- "Material Health": Materials must be verified as healthy and safe for humans and the environment.
- "Design for Disassembly, Reuse, and Recovery": Design materials, products, and components so they can be disassembled during demolition or renovation, retain their full value, and be returned to the supply chain for use in new installation or construction.
- "Value Chain Collaboration": Developing lines of communication across the supply chain to emphasize the reuse of materials including between manufacturers, designers, owners, banks and financial institutions to facilitate the discussion of how to incorporate reused materials.
- "Realizing Value": Retain or create value over the use and reuse of a product. Building Information Modeling (BIM) aids this process through analyzing and storing data on materials and components, allowing for identification, optimization, verification, and tracking of materials and components.
- "Policy and Standards": Governmental agencies, local, regional, and state governments, can promote the circular economy through incentive, policy, law, and voluntary standards. Agencies and governments have a large impact since they provide control and oversight of solid waste disposal.

4.4.2 Biodegradability

Biodegradation is a chemical process where microorganisms break down a material into carbon dioxide, water, and naturally occurring minerals. For a material to be biodegradable, complete decomposition must occur in a year or less. Biodegradable products may be either solid or liquid. Biodegradable building materials are rare because biodegradability often conflicts with durability, weatherproofing, and other performance requirements [27, 28].

During decomposition, materials cannot produce toxic substances. Often biodegradable materials require oxygen for decomposition and may not decompose in oxygen-deprived systems such as a sewage treatment plant or landfill. Additionally, the release of biodegradable material into the environment must balance with the availability of microorganisms. Too much biodegradable material left in a place can overwhelm and damage an ecosystem locally or regionally.

In order to claim a product is biodegradable, infrastructure must be in place at the location that the product is discarded to allow biodegradation. If this infrastructure is not available, then the labeling is considered false. The FTC has brought litigation against biodegradable claims, citing the Federal Trade Commission Act, 15 U.S.C. § 45, prohibiting deceptive advertising.

4.4.3 Compostable

Compostable materials are solid products that can biodegrade under certain conditions, such as a certain temperature or within a certain time frame [27, 28].

The expectation is that composted materials will decompose at the same rate as the materials with which they are composted (often lawn clippings, leaves, or vegetable matter). Often these conditions are more specific than what landfills offer and are provided by a specifically designated municipal composting facility. Compostable is similar to biodegradable, but instead of leaving no trace, organic nutrient-rich materials remain at the site of decomposition. A composting site is accessible, allowing for retrieval of the nutrient-rich material, such as mulch.

There is no federal regulation for composting, but the FTC has brought litigation against compostable claims, citing the Federal Trade Commission Act, 15 U.S.C. § 45, prohibiting deceptive advertising. Deceptive advertising or claims about sustainable initiatives are considered "green washing."

4.4.4 Recyclability

A recyclable material can be reprocessed or dismantled to be reused. To be recyclable, conditions for a product include the following: the material must have a market value for reuse and can return to the supply chain as a component or feedstock, and an existing collection, sorting, and processing system by waste processors must have enough robustness that the material will likely be recycled. The FTC actively pursues legal action against companies that provide misleading claims about recyclability. In particular, the FTC focuses upon the difference between what can technically be recycled and what is likely to be recycled.

A product made of recycled or recyclable materials is not necessarily recyclable. For example, a cup made of recyclable paper, coated in recyclable plastic, fuses two recyclable materials together. Because they cannot easily be separated, the product is no longer recyclable. Coatings and adhesives for building products make recyclability particularly difficult. To that end, one of the tenets of the circular economy is to emphasize mechanical fasteners that can be unscrewed over the use of adhesives. For the recycled plastics industry, baseline conditions by which a product can be defined as recyclable have been developed by Plastics Recyclers Europe and the Association of Plastic Recyclers [29].

Product Packaging

Life cycle assessment began with comparison of packaging back in the 1970s as industries realized the large environmental impact associated with packaging. Packaging serves many purposes:

- Relaying quality and brand
- Ensuring hygiene and safety
- Maintaining security
- Providing convenience of storage and the ability to buy in bulk quantity
- For manufacturers, providing ease of transportation and storage

Metal cans, plastic containers, cardboard boxes, paper, and plastic film are used for packaging today. The range of initiatives to decrease the environmental impact of packaging demonstrates a large number of the strategies described in this chapter:

- Ban single-use packaging. A good example is the current ban of single-use plastic grocery bags in a variety of countries and across certain states in the United States.
- Design packaging for reuse and establish a system that ensures reuse.
- Design packages without red list chemicals (lists of toxic chemicals used in products and pharmaceuticals) or restricted substance lists (lists used by the consumer product, apparel, and shoe industry).
- Develop a method that ensures that packaging will be recycled or composted by municipal waste centers.
- Design packaging that is recyclable or compostable.
- Develop standards for manufacturers to measure the amount of packaging they use, publicly disclose the resulting data, and after public disclosure, set goals of reduction, limits, and bans on packaging.

4.4.5 Design for Disassembly

Building materials designed for disassembly can be cost-effectively and rapidly demounted and removed from their original location, and then reassembled, and reinstalled in a new location [30]. Design for disassembly may refer to the design of an entire structural system or to a single product, like a chair. Products designed for disassembly can do the following:

- Conserve resources making it easier to extend the life of the product through reuse as individual parts or through easy replacement of individual parts that may break during a product's useful life.
- Decrease installation time and labor.
- Avoid the use of toxic glues and adhesives by prioritizing mechanical fasteners.

To accomplish design for disassembly, the following is useful:

- Planning for disassembly is documented often through BIM, with information for reuse, disassembly, and reinstallation of components.
- Use of standard-sized modular components.
- Durability of materials and components that maintain value over time.
- Provision for visible and accessible connections.
- Use of mechanical fasteners in lieu of adhesives.

4.4.6 *Manufacturing Waste Recovery and Resource Conservation*

During the processing and manufacturing stages of a product's life, there are many opportunities for recovering waste. The EPA regulates waste recovery and reduction through the RCRA [31]. This law includes:

- Identification of waste types.
- Transportation, sorting, and disposal methods for each waste type.
- Discussion of how waste should be handled by waste management companies and governments. This includes methods for adhering to the EPA waste hierarchy pyramid, from preparing materials for recycling to incineration or encapsulation in landfills.

Individual states and municipalities further define waste recovery and reduction efforts through local legislation [32].

Manufacturing facilities that divert 90% or more of their waste from landfill through recycling or recovery and achieve other waste reduction strategies may achieve zero-waste certification. TRUE (Total Resource Use and Efficiency) is a certification program that may be pursued by manufacturers and others throughout the product life cycle to define and achieve zero waste [33]. The certification is administered by Green Business Certification Inc (GBCI). Other certification bodies may certify facilities as zero waste as a single attribute certification. Additional information on certifications is presented in Chap. 10.

4.4.7 *Material Recovery and Manufacturer "Take-Back" Programs*

After a product has reached the end of its useful life, a manufacturer may collect material from the consumer to be recycled into new products (Table 4.7). Legislation and regulation that initiates zero-waste municipalities and charges or fines for not collecting materials will motivate manufacturers to develop more comprehensive recovery and reuse programs in the future. A few examples of material recovery initiatives include take-back programs, the concept of extended producer responsibility, and materials recovery facilities.

Table 4.7 Waste recovery initiatives

Initiative	Description	Materials
Take-Back Program	Manufacturer may collect material from the consumer to be recycled into new products In theory, enables the manufacturer to accurately calculate post-consumer recycling percentages In practice, low amounts of materials are actually returned through these programs	Carpet tiles Furniture Lamps
Extended Producer Responsibility	Manufacturers should be responsible for a product throughout its entire life cycle, rather than only up until the point of sale Encourages sustainable ingredient selection and waste recovery practices by putting the financial burden of recycling onto manufacturers Manufacturers are less motivated by the initial cost of ingredients, rather than the ability to recycle those ingredients into new feedstocks at the end of their useful life	Building product and furniture rentals. This reduces the risk of products and furniture ending up in landfill and increases the motivation for manufacturers to build durable, reusable products to extend the financial life of the rental product [34]
Materials Recovery Facility (MRF)	MRFs adhere to the RCRA An MRF can calculate the amount of waste a manufacturer or construction site creates and the amount of waste and recyclable material consumers/project owners generate	Collect materials discarded during manufacture, installation, and demolition

References

1. LEED User. (2020). *A note of caution on regional materials in LEED v4*. Building Green. https://leeduser.buildinggreen.com/content/note-caution-regional-materials-leed-v4. Accessed 3 Sept 2020.
2. International Living Future Institute. (2019). *Living building challenge 4.0*. International Living Building Institute, International Living Future Institute and Cascadia Green Building Council, p. 55. https://www2.living-future.org/LBC4.0?RD_Scheduler=LBC4. Accessed 3 Sept 2020.
3. Source Intelligence. (2020). *What are conflict minerals?* https://www.sourceintelligence.com/what-are-conflict-minerals/. Accessed 1 Sept 2020.
4. United States Department of Agriculture. (2020). *What is Biopreferred?* https://www.biopreferred.gov/BioPreferred/faces/pages/AboutBioPreferred.xhtml. Accessed 9 Sept 2020.
5. United States Green Building Council. (2020). LEED *BD+C LEED v4 Environmentally preferable products materials and resources*. https://www.usgbc.org/credits/mid-rise/v4-draft/mrc2 . Accessed 10 Sept 2020.

6. Forest Stewardship Council. (2012). *Forest Stewardship Council vs. Sustainable forestry initiative a comparison of the standards*. https://www.nrcm.org/wp-content/uploads/2013/09/FSCvSFIstandards.pdf. Accessed 9 Sept 2020.
7. Forest Stewardship Council. (2015). *US Federal procurement recommends FSC Certification*. https://fsc.org/en/news/us-federal-procurement-recommends-fsc-certification. Accessed 9 Sept 2020.
8. Perkins, A. (2019). *Earning LEED points with certified wood*. USGBC LEED. https://www.usgbc.org/articles/earning-leed-points-certified-wood . Accessed 9 Sept 2020.
9. Sustainable Forestry Initiative Program. (2016). *Government Leaders recognize the value of SFI program for sustainable forest management and responsible procurement*. SFI News Release. https://www.sfiprogram.org/government-leaders-recognize-the-value-of-sfi-program-for-sustainable-forest-management-and-responsible-procurement/. Accessed 9 Sept 2020.
10. Andaya, Romy "Skip". (2019). *Spanish Moss, Cypress, and Henry Ford, oral history*.
11. Grey, J. (2019). *Reclaimed materials. sustainable build*. http://www.sustainablebuild.co.uk/reclaimedmaterials.html. Accessed 9 Sept 2020.
12. Life Magazine. (1955). Throwaway Living. *Life Magazine*, pp. 43–44, Vol. 39, #5, Aug 1, 1955. https://books.google.com/books?id=xlYEAAAAMBAJ&printsec=frontcover#v=onepage&q&f=false. Accessed 10 Sept 2020.
13. The Recycling Partnership. (2020). *2020 State of curbside recycling report*, p. 8. https://recyclingpartnership.org/wp-content/uploads/dlm_uploads/2020/02/2020-State-of-Curbside-Recycling.pdf. Accessed 9 Sept 2020.
14. Public Law 94-580 94th Congress. (1976). *Resource conservation and 42 USC 6901 Recovery Act of 1976*. https://www.govinfo.gov/content/pkg/STATUTE-90/pdf/STATUTE-90-Pg2795.pdf. Accessed 9 Sept 2020.
15. Robert, T. (2008). *Defining recycled content*. Building Green, Vol. 17, Issue 12. https://www.buildinggreen.com/primer/defining-recycled-content. Accessed 9 Sept 2020.
16. United States Environmental Protection Agency. (2017). *National overview: Facts and figures on materials, wastes and recycling*. https://www.epa.gov/facts-and-figures-about-materials-waste-and-recycling/national-overview-facts-and-figures-materials. Accessed 9 Sept 2020.
17. Trickey, E. (2019). San Francisco's quest to make landfills obsolete. *Politico Magazine*. https://www.politico.com/news/magazine/2019/11/21/san-francisco-recycling-sustainability-trash-landfills-070075. Accessed 9 Sept 2020.
18. SF Department of the Environment. (2003). *Resolution setting zero waste date*. Resolution No. 002–03-COE. https://sfenvironment.org/sites/default/files/editor-uploads/zero_waste/pdf/resolutionzerowastedate.pdf. Accessed 9 Sept 2020.
19. SF Department of the Environment. (2018). *Mayor London breed challenges cities, states and regions around the world to join san francisco in setting aggressive sustainability goals*. https://sfenvironment.org/press-release/mayor-london-breed-challenges-cities-states-and-regions-around-the-world-to-join-san-francisco-in-setting-aggressive-sustainability. Accessed 9 Sept 2020.
20. SF Department of the Environment. (2006). *Construction and demolition debris ordinance*. City Ordinance No. 27–06. https://sfenvironment.org/article/other-local-sustainable-buildings-policies/construction-and-demolition-debris-ordinance. Accessed 9 Sept 2020.
21. Scholand, M., Dillon, H. (2012) *Life-cycle assessment of energy and environmental impacts of LED lighting products part 2: LED manufacturing and performance*. Pacific Northwest National Laboratory N14 Energy Limited. https://www1.eere.energy.gov/buildings/publications/pdfs/ssl/2012_led_lca-pt2.pdf. Accessed 9 Sept 2020.
22. United States Environmental Protection Agency. (2018). *Sustainable materials management: Non-Hazardous materials and waste management hierarchy*. https://www.epa.gov/smm/sustainable-materials-management-non-hazardous-materials-and-waste-management-hierarchy. Accessed 9 Sept 2020.
23. McDonough, W. (2002). *Cradle to cradle*. https://mcdonough.com/cradle-to-cradle/. Accessed 11 Sept 2020.

24. Ellen Macarthur Foundation. (2017). *Circular economy*. https://www.ellenmacarthurfoundation.org/. Accessed 11 Sept 2020.
25. Terracycle. (2020). *Terracycle*. https://www.terracycle.com/en-US/. Accessed 11 Sept 2020.
26. Cradle2Cradle. (2017). *Green grows up: Building in the age of the circular economy*. https://www.c2ccertified.org/news/article/green-grows-up-building-in-the-age-of-the-circular-economy. Accessed 11 Sept 2020.
27. European Bioplastics. (2016). *What is the difference between 'biodegradable' and 'compostable'?* https://www.european-bioplastics.org/faq-items/what-is-the-difference-between-biodegradable-and-compostable/. Accessed 11 Sept 2020.
28. Greengood. (2013). *Biodegradable and compostable definitions*. http://www.greengood.com/terms_to_know/biodegradable_and_compostable_definitions.htm. Accessed 11 Sept 2020.
29. Messenger, B. (2018). *Global definition of plastics recyclability from international recycling associations*. Waste Management World. https://waste-management-world.com/a/global-definition-of-plastics-recyclability-from-international-recycling-associations. Accessed 11 Sept 2020.
30. United States Environmental Protection Agency. (2020). *Best practices for reducing, reusing, and recycling construction and demolition materials*. https://www.epa.gov/smm/best-practices-reducing-reusing-and-recycling-construction-and-demolition-materials. Accessed 11 Sept 2020.
31. United States Environmental Protection Agency. (2020). *Resource Conservation and Recovery Act (RCRA) laws and regulations*. https://www.epa.gov/rcra. Accessed 11 Sept 2020.
32. CalRecycle. (2019). *CalRecycle enforcement and compliance programs*. https://www.calrecycle.ca.gov/enforcement. Accessed 11 Sept 2020.
33. GBCI. (2020). *True program zero waste certification*. https://true.gbci.org/. Accessed 11 Sept 2020.
34. Lieber, C. (2019). *Ikea will soon offer furniture rentals because the end of ownership is near*. The goods by Vox. https://www.vox.com/the-goods/2019/2/5/18212518/ikea-furniture-rentals-sharing-economy. Accessed 11 Sept 2020.

Chapter 5
Energy Use

5.1 Introduction

Energy is used at every stage in a product's life cycle, from collecting raw ingredients, to processing and manufacturing, to use, and finally at the end of life (Fig. 5.1). Transporting materials and products between life cycle stages also requires energy in the form of fuel (Fig. 5.2).

To reduce the volume of energy consumed and the impacts resulting from energy use, there are several key questions to consider:

- What is the energy source? Can a more sustainable source be used?
- Is it possible to consume less energy across the life cycle?
- Can excess or potential energy be recovered during one or multiple life cycle stages?

To aid in prioritizing various strategies for reducing energy use, Peter Wolfe, the former Director General of the Renewable Energy Association established the following hierarchy (Fig. 5.3) [1]:

- Energy savings by turning equipment off, eliminating waste
- Energy efficiency by replacing appliances with ones that have lower energy loss
- Using renewable fuels
- Reducing carbon emissions and/or capturing carbon
- Purchasing carbon offsets when using conventional fuels to financially support increasing renewable energy sources across the energy grid

H. R. Roth et al., *The Green Building Materials Manual*,
https://doi.org/10.1007/978-3-030-64888-6_5

INITIATIVES AND IMPACTS: ENERGY USE

Energy

Energy Source: Energy Fuel vs Flow
Renewable Energy: Grid, On Site, & Exported
Energy Savings and Efficiency: Energy Consumption

Audits
Energy Efficiency
Excess and Potential Energy: Energy Recovery
Energy from Waste
Bio-energy from Waste

Fig. 5.1 Initiatives and impacts for energy use

ENERGY USE DURING LIFE CYCLE

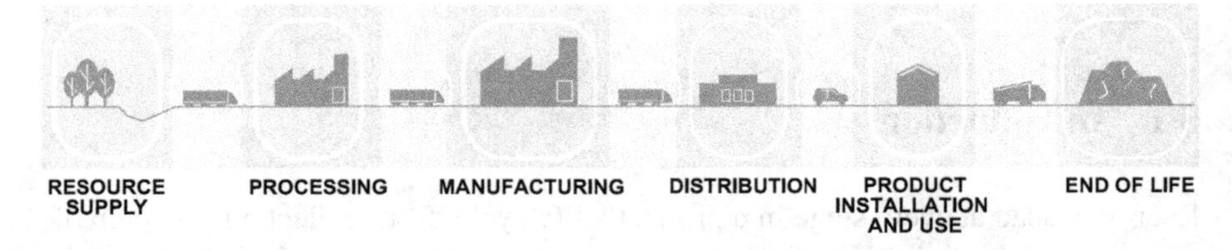

Fig. 5.2 Energy is used during every stage of life cycle and every method of transportation between stages

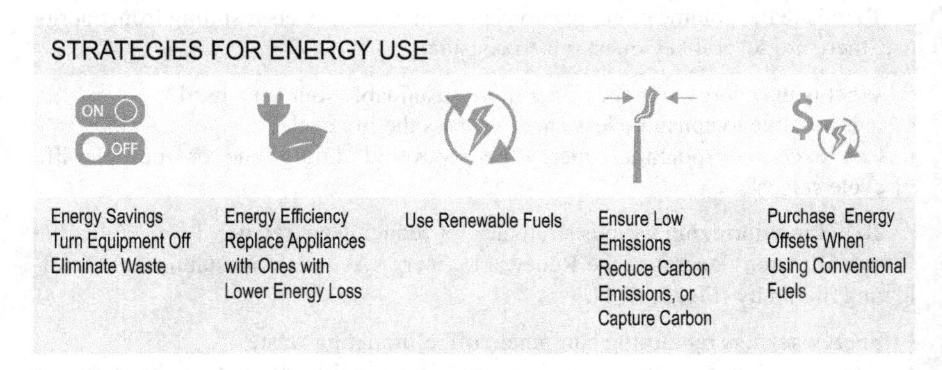

Fig. 5.3 Strategies for energy use

5.2 Energy Sources

Energy is derived from raw materials or renewable energy flows that are converted into fuel or electricity. Fuel and electricity are then distributed to equipment, facilities, and vehicles at every stage in a product's life.

The energy sector uses specific terminology to describe the character of different energy sources (Fig. 5.4) [2–5]:

- Primary energy fuel or source
- Secondary energy fuel or source
- Energy carrier

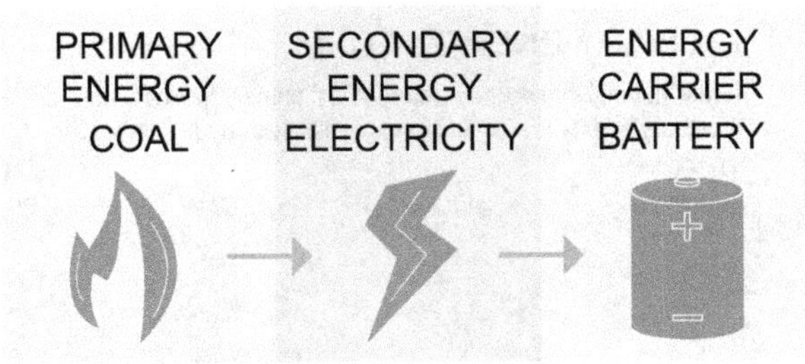

Fig. 5.4 Energy sources

5.2.1 Primary Energy Fuel

A primary energy fuel is a raw material that is unconverted. Most primary energy fuels are non-renewable. A primary energy fuel is consumed, typically through burning. Usually, it is consumed at a plant, where it is often made into a secondary transportable fuel. Primary energy fuels include:

- Petroleum
- Natural gas
- Coal
- Burnable waste
- Nuclear energy

A primary energy flow is not consumed but instead flows past a device that collects its energy. Renewable energy is an interchangeable term with primary energy flow and is discussed in further detail later in this chapter. Common primary energy flows include:

- Flowing water (hydroelectric)
- Wind (wind turbines)
- Solar radiation (solar collector)
- Geothermal sourcing and materials that give off heat as they decompose, if that heat is collected
- Tidal energy

Like a primary fuel, a primary energy flow is typically converted into a transportable secondary energy fuel or currency (Fig. 5.5).

5.2.2 Secondary Energy Fuel or Energy Currency

A secondary energy fuel is captured or made from a primary fuel. Gasoline is distilled from petroleum. Methane is also distilled from petroleum or captured from biological waste off-gassing. A secondary energy fuel typically can be delivered to a motor that uses the energy. The motor is often at a facility or in a transport vehicle. Secondary energy can also be converted into electricity. Examples include:

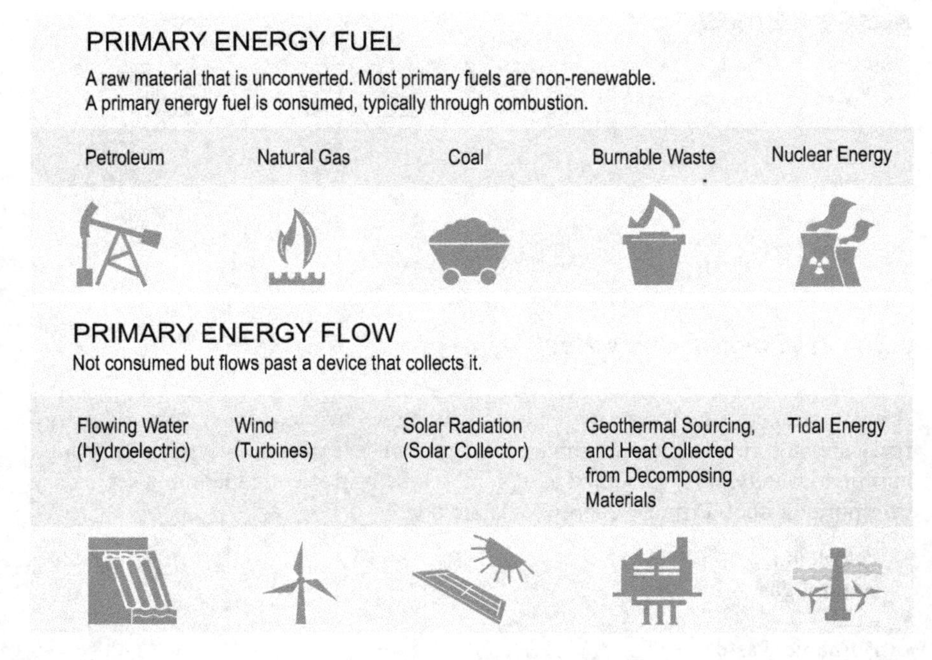

Fig. 5.5 Primary energy fuels and primary energy flow

- Gasoline
- Methane
- Hydrogen
- Natural gas

An energy currency can be carried and delivered using several methods:

- Gasoline, methane, and natural gas are easily transported, stored, and used directly by combustion engines.
- Hydrogen is stored in fuel cells.
- Electricity is transported across the power grid after it is created by burning fuel or converting primary/renewable energy flows.

5.2.3 *Energy Carrier*

An energy carrier is a material or system that has potential or holds energy for later use:

- Raw materials like wood, coal, petroleum, and natural gas
- Damned water, pressurized air

- Hydrogen and other materials used for chemical or nuclear reaction
- Batteries and capacitors

5.2.4 Renewable Energy

Renewable energy refers to energy derived from sources that naturally replenish. These sources may vary in availability depending on time of day or year and geographic location. They do not always deliver a constant amount of energy. To maintain a consistent supply to users, these energy sources sometimes are augmented with other renewable energy sources or fossil fuels. Often renewable energy sources can be described as primary energy flows, as described in Sect. 5.2.1.

Organizations and governments disagree on what energy sources are truly renewable and which should be considered to be low in environmental impact. In the United States, the Energy Information Administration (EIA) lists five sources of renewable energy:

1. Biomass, including wood and wood waste, municipal solid waste, landfill gas and biogas, ethanol, and biodiesel
2. Hydropower
3. Geothermal
4. Wind
5. Solar

In 2019, coal and renewable energy each accounted for 11% of the US energy consumption. Coal consumption decreased 9% since 2018 (from 11.73%) [6]. Within renewable energy, the five resources were distributed in the following way in 2019 [7]:

- 45% of that volume of energy was from biomass sources (burned).
- 22% from hydropower.
- 24% from wind.
- 9% from solar.
- 2% from geothermal (Fig. 5.6).

Energy sources produce emissions over their life, including carbon dioxide and other greenhouse gases (GHGs). Renewable energy sources are key to reducing the release of GHG emissions into the air to prevent climate change (see Chap. 7 for a more detailed discussion of greenhouse gases and climate change).

Changing the energy source used to extract or manufacture a product to a lower carbon footprint fuel may result in reducing the overall carbon footprint of a product without reducing the total quantity of energy used. For example, running an appliance for an hour using electricity from solar panels has a lower environmental impact than using the same appliance for an hour while using electricity sourced from coal.

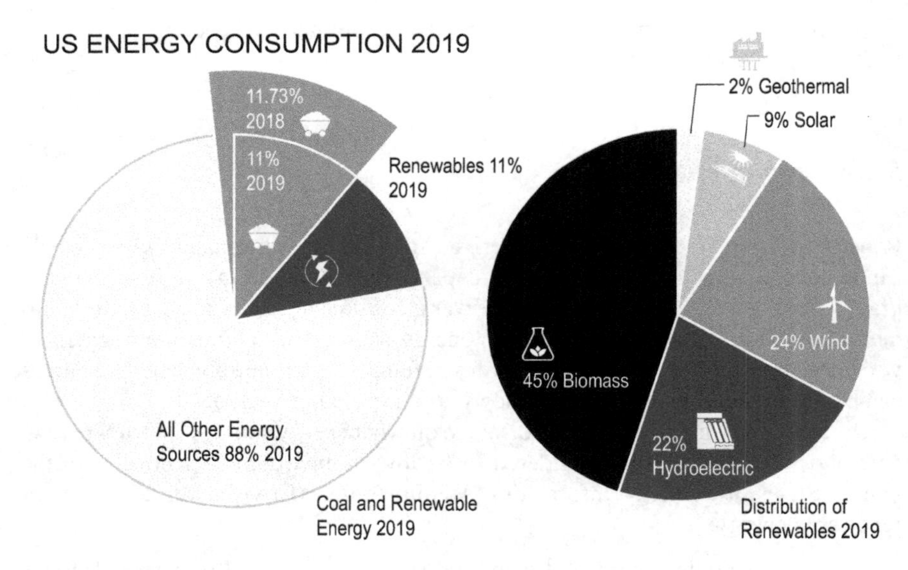

Fig. 5.6 US energy consumption in 2019, with the distribution of energy sources

While renewable energy sources have fewer environmental impacts than fossil fuel-derived energy sources, they are not without environmental tradeoffs:

- Hydroelectricity can negatively impact aquatic habitats, including fish populations.
- Solar energy requires large tracts of land in specific geographic locations to maximize its efficacy, where it may compete with other land uses such as agriculture or housing.
- Solar collectors use rare earth metals, which may be conflict resources.
- Biomass creates emissions when burned.

Certifications often reward energy efficiency or use of renewable energy during the life cycle of a product or building. Energy efficiency can be accomplished through a certification requiring limits or reductions of the total amount of fuel used during manufacturing or building occupation. While regulations and standards will not typically require the use of a specific type of energy source, they may limit the emissions of a facility or product. This results in a limitation on what fuels can be used.

5.2.5 *On-Site Renewable Energy*

On-site solar collectors, wind turbines, and methods of harnessing geothermal flow provide renewable energy directly to a building on-site. After the cost of equipment, building owners can collect energy free of charge with a source of energy that is

Table 5.1 Weighing the positives and negatives of on-site renewable energy

Location	Regional Conditions	Relative Impact of On-site Renewables
Midwestern City	Coal-based power grid	*On-site is a positive choice:* On-site solar collection produces no emissions compared to the high emissions of the power grid
Seattle	Little direct sunlight Hydroelectric-based power grid	*On-site is a toss-up or bad choice:* On-site solar collection is inefficient, creating less impact, especially since an abundant renewable resource fuels the municipal power grid
San Francisco	Regulations in place to provide carbon-neutral power grid	*On-site is a toss-up or bad choice:* On-site solar collection produces no emissions, but the local power grid produces no emissions also

emissions-free. When on-site equipment cannot meet an entire energy need, a facility can use energy from the broader municipal grid to fulfill the need.

The contribution of on-site renewable energy has a relative impact in relation to the municipal electrical grid. The impact of providing on-site renewable energy varies depending on (Table 5.1):

- Geographic location
- Climatic factors
- Regulatory environment for the region

5.2.6 Exported Energy

Sometimes on-site devices, such as solar collectors or wind turbines, produce more energy than is required by a facility. In those cases, the devices send the excess supply to the power grid. The local utility pays the facility for this energy. If the energy is produced using renewable energy, manufacturers can support the infrastructure by purchasing carbon offsets.

Carpet Companies and Solar Collectors: Quantifying the Narrative

Many carpet industry manufacturers primarily focus their sustainability efforts on recycling, water use, social accountability, and energy in lieu of toxicity and emissions. This is no surprise, as the raw materials for carpet are often petroleum-based.

Companies such as Mohawk Industries publish statements on sustainability annually. Mohawk Industries has developed solar arrays at several of its plants as an environmental initiative. In 2015, Mohawk announced the implementation of a 1400 panel array in Summerville, Georgia. Mohawk states that this array added energy to the electric grid that would power about 40 homes. Additionally, this array brought Mohawk's total nationwide production to 250 megawatts of solar energy per year [8]. 250 MW seems a substantial number. This number represented the total capacity of generation. However, solar panels operate at an efficiency rate as low as 25% capacity [9]. This lower threshold means that Mohawk could generate a minimum electricity equivalent to approximately 6000 homes [10]. This is still a significant number.

However, Mohawk does not publicly relate this production of solar energy to their own usage of energy during their manufacturing processes. Mohawk states in their 2017 sustainability report that they have achieved a 5.61% reduction in energy intensity since 2010 [11]. It is good that Mohawk is making efforts to decrease their energy usage, but a percentage in reduction does not provide an actual quantity of energy used per year. As a result, Mohawk's energy usage cannot be compared to any other competitor company's usage. This situation is not specific to Mohawk: many manufacturers across all product sectors disclose energy use in terms of percentage reduction rather than specific numbers.

5.3 Energy Savings and Efficiency

Reducing the environmental impacts of energy use can be accomplished across a product's life cycle through:

- An absolute reduction in energy use, like turning off lights during the day
- An increase in efficiency, like replacing incandescent light bulbs with LEDs

For building products, the largest opportunities for energy savings and efficiency are often in the manufacturing stage of the life cycle while the product is at a processing or manufacturing plant.

Additionally, new appliances and equipment can use less energy or be more efficient during operation or use, once they are installed. The voluntary Energy Star Program, administered by the EPA, persuades manufacturers to increase appliance and equipment efficiency, and requires publication of energy consumption data for consumer comparison. Federal and state government purchasing programs require that equipment procured by government agencies bear the Energy Star mark. Because agencies are such large purchasers, this encourages manufacturers to adhere to the Energy Star's ever-increasing standards.

At the scale of a whole building, systems can also reduce energy use. For example, building automation systems can turn off equipment or lighting when not being used.

5.3.1 Energy Consumption

Energy consumption is the amount of energy used by machinery, appliances, digital devices, systems, and products. Energy consumption can be considered at the product level, across a region, a country, or the globe.

The amount of energy consumed is different from the amount of primary energy used. Energy consumed includes the amount of energy that is used to refine raw materials, to convert between energy types, and to move energy from one location to another across the energy grid. Fundamentally each of these stages has some inefficiency (Table 5.2).

The EIA publishes data on world energy consumption. While there are assumptions made in data modeling, this published data provides a relative comparison between countries and across the globe over time revealing trends and changes [13].

5.3.2 Energy Audits

An energy audit analyzes energy use across a building or facility to measure consumption and evaluate options to increase the building's efficiency. Residential energy audits can be a low-cost, easy step to improving energy efficiency, particularly for older homes. Local governments often subsidize residential energy audits or even provide them for free, whereas commercial energy audits such as those for a manufacturing facility are more extensive.

Key components of an energy audit include analysis of the building's utility bill, a visit to the building to survey and calibrate the building's lighting systems, heating, ventilation, and air conditioning (HVAC) equipment, and other major energy-consuming systems and equipment. The survey also includes evaluating the

Table 5.2 Energy is consumed as it transfers from one life cycle stage to another. This consumption causes inefficiency

Stage	Loss or Inefficiency
Refine	Energy is lost during the extraction and refinement of raw materials in preparation to use them as primary fuel. Energy is also lost while collecting and concentrating primary energy flows through equipment. For instance, a stationary solar collector will not be at the optimal angle to collect solar energy as the sun moves across the sky
Convert	Energy is lost converting between primary and secondary energy sources. For example, distilling petroleum to become gasoline or burning coal to become electricity requires energy. No conversion process is 100% efficient
Transport	Energy can be lost as a secondary fuel through leaks or evaporation. Up to 5% of electricity is lost during transmission across the energy grid [12]

building's exterior envelope, as this has a large impact on the building's overall efficiency.

The American Society of Heating, Refrigerating and Air-Conditioning Engineers (ASHRAE) defines three levels of energy audits, ranging from a quick identification of low- or no-cost energy efficiency opportunities to an extensive analysis of deep energy retrofits:

- Level 1 – Walk-Through Assessment
- Level 2 – Energy Survey and Analysis
- Level 3 – Detailed Analysis of Capital-Intensive Modification

5.3.3 Energy Efficiency

Energy efficiency has grown to mean several things. Depending on how the term is used, it can scale from how much energy a single light bulb uses to the goals of an entire company's operations (Table 5.3).

Glass Manufacturers: Efficiency of Product, Inefficiency of Process

The glass and curtain wall industry emphasizes energy efficiency of window systems once they are installed in a building. This focus originated in the energy crises of the 1970s. Energy efficiency calculations are established by ASTM standards. With nearly 50 years of data, the industry has become very competitive and technologically advanced in the measurement of resistance (R Value) and transmittance (U-factor and SHGC) of energy.

Corporate disclosure of energy use during manufacturing provides less specificity. Oldcastle BuildingEnvelope is of particular interest. In 2018 Oldcastle presented a leadership position on sustainability that aligned with LEED credits [14]. Since then that position has been deleted, and the company website remains silent on sustainable initiatives.

By contrast, Pilkington publishes a statement on energy usage that provides a broader approach, citing information on the current landscape of energy supply and how little is generated by renewable resources. Pilkington's efforts focus on energy offsets and some heat recovery. While they specifically quote that the company received "8.7 GWh from renewables and waste heat directly, plus a further 38 GWh from combined heat and power" [15], nowhere do they relate that number to their overall expenditure of energy. For reference, heat recovery systems are typically 50–80% efficient [16]. The company also publishes information on approximate CO_2 emissions values for their float glass during the manufacturing process.

Table 5.3 Different ways to define energy efficiency

Type of Efficiency	Description
Efficiency and Energy Loss	Energy loss can be through heat, light, or vibration. A motor may vibrate heavily and give off a tremendous amount of heat; as a result, it has decreased efficiency Power output/power input = efficiency Many manufacturers seek to increase the efficiency of equipment and appliances
Effectiveness During Use	A light bulb is efficient if it is effective at producing visible light tuned to the sensitivity of the human eye This type of efficiency can also be defined as "efficacy"
Passive Transfer of Energy	Efficiency can be calculated by measuring across a system that provides a passive barrier to energy transfer. For example, insulation or energy-efficient windows contribute to the energy efficiency of a home Variables measuring energy efficiency include: *R value*, the resistance to heat flow through an individual product, such as an insulated metal panel *U-factor*, the ability to conduct heat across an assembly, taking into account the entire system, such as a window assembly *Solar heat gain coefficient* (SHGC), typically for glass measuring its ability to reduce heat gain by blocking the sun's radiation *Air infiltration*, which measures how well sealed an assembly is
Goal of Reducing Energy Usage	Often this is measured in terms of the percentage decrease in the use of fuels, use of electricity, or a goal of decreasing carbon footprint

5.4 Excess and Potential Energy Throughout Life Cycle

The processing and manufacturing stages of life cycle are the predominant stages using energy and materials. Processing and manufacturing do not use energy with 100% efficiency. Excess energy, such as heat, is often released into the environment. At the same time, processing and manufacturing produce excess materials as waste. For example, planed lumber results in sawdust and wood shavings. There are several ways to use this excess energy and excess waste to produce energy.

5.4.1 Energy Recovery

Throughout the life cycle, facilities use mechanical processes and chemical reactions that generate extra energy. This excess energy comes in the form of:

- Heat
- Air or steam pressure
- Excess chemicals, many of which can be considered combustible fuels

These sources of energy can be recovered and transferred into heat or an energy currency such as electricity, which can either be used on-site for facility operations or returned to the energy grid [17].

5.4.2 Energy from Waste

Material waste can be converted to produce heat, electricity, or a secondary fuel similar to gasoline (Table 5.4). Incineration of waste or conversion to a fuel that is later burned results in emissions. However, heat recovery from waste does not necessarily produce emissions.

Municipal Waste

In a municipal environment, a city's waste and trash can be reduced through incineration. Historically almost all cultures burned their waste. With the advent of the Industrial Revolution, incineration increased, particularly of toxic materials. In an attempt to control widespread burning, New York City built its first publicly owned incinerator in 1885. In 1970 the Clean Air Act placed restrictions on incineration emissions, which were previously uncontrolled. In the 1990s the EPA enacted Maximum Achievable Control Technologies regulations, further controlling emissions.

According to the EPA, municipal solid waste (MSW) in 2017 (the most recent statistics) was 267.8 million tons [19]:

- 35% was either recycled or composted.
- 13% was combusted with energy recovery.
- 52% was deposited in landfills.

5.4.3 Bioenergy from Waste

Biomass is a biological material used specifically for energy production [20]. Biomass is made of a living material. It can be:

Table 5.4 Ways in which waste can be converted into energy [18]

Process	Description
Combustion	Burning waste
Gasification	Exposing a waste material to oxygen at a high temperature creates a gas that is combustible. If the waste material is biological, this is considered a renewable energy source
Pyrolization	Chemically changing a material at a high temperature to make it into a fuel that is more effective or efficient. Often these resultant fuels produce high emissions, for example, making wood into charcoal or coal into coke. Biological materials can be turned into biochar, and recycled plastics can be returned to oil-based fuels
Digestion	Anaerobic digestion breaks down biological material to produce biogas, which can be burned
Chemical Reaction	Chemical reactions can produce heat or fuel as a byproduct. These byproducts vary in toxicity

- Waste after harvesting of crops or timber
- Food waste
- Rapidly renewable crops such as corn
- Microalgae

Bioenergy can be produced by distilling fuel from biomass, through burning, or by capturing energy produced by biological decay. When bioenergy is converted to an energy currency, it behaves like gasoline, jet fuel, or diesel and is easily transported. Burning biomass generates emissions.

References

1. Wolf, P. (2005). *A proposed energy hierarchy. Wolfeware*. https://en.wikipedia.org/wiki/Energy_hierarchy#Origins_of_the_energy_hierarchy. Accessed 12 Sept 2020.
2. Donev, J. M. K. C., et al. (2020). *Energy education – Primary fuel*. University of Calgary. https://energyeducation.ca/encyclopedia/Primary_fuel. Accessed 12 Sept 2020.
3. Donev, J. M. K. C., et al. (2020). *Energy education – Secondary fuel*. University of Calgary. https://energyeducation.ca/encyclopedia/Secondary_fuel. Accessed 12 Sept 2020.
4. Donev, J. M. K. C., et al. (2020). *Energy education – Energy currency*. University of Calgary. https://energyeducation.ca/encyclopedia/Energy_currency. Accessed 12 Sept 2020.
5. Satayapal, S. (2020). *Hydrogen: A clean, flexible, energy carrier*. US Department of Energy. https://www.energy.gov/eere/articles/hydrogen-clean-flexible-energy-carrier. Accessed 12 Sept 2020.
6. Francis, M. (2020). *U.S. Renewable Energy consumption surpasses coal for the first time in over 130 years*. Today in Energy, US Energy Information Administration. https://www.eia.gov/todayinenergy/detail.php?id=43895. Accessed 12 Sept 2020.
7. US Energy Information Administration. (2020). Renewable energy production and consumption by source. *Monthly Energy Review* August 2020. https://www.eia.gov/totalenergy/data/monthly/pdf/sec10_3.pdf. Accessed 12 Sept 2020.
8. The Chattanoogan. (2015). *Mohawk installs new solar array at summerville facility*. https://www.chattanoogan.com/2015/4/2/297344/Mohawk-Installs-New-Solar-Array-At.aspx. Accessed 12 Sept 2020.
9. United States Nuclear Regulatory Association. (2012). *What is a megawatt?* p. 3. https://www.nrc.gov/docs/ML1209/ML120960701.pdf. Accessed 12 Sept 2020.
10. U.S. Energy Information Administration. (2019). *How much electricity does an American home use?* Frequently Asked Questions. https://www.eia.gov/tools/faqs/faq.php?id=97&t=3. Accessed 12 Sept 2020.
11. Mohawk Industries. (2017). *2017 Sustainability report, believe in better*, p. 18. https://mohawkind.com/_pdf/Mohawk_2017_Sustanability_Report.pdf. Accessed 12 Sept 2020.
12. U.S. Energy Information Administration. (2019). *How much electricity is lost in electricity transmission and distribution in the United States?* Frequently Asked Questions. https://www.eia.gov/tools/faqs/faq.php?id=105&t=3. Accessed 12 Sept 2020.
13. U.S. Energy Information Administration. (2017). *Rankings about energy in the World*. https://www.eia.gov/international/overview/world. Accessed 12 Sept 2020.
14. Old Castle Building Envelope. (2018). *Leadership position*. https://obe.com/sustainability. Accessed 01 Nov 2018, No Longer Accessible.
15. Pilkington. (2020). *Frequently asked questions about sustainability*. https://www.pilkington.com/en/global/commercial-applications/sustainability/sustainability-faq. Accessed 12 Sept 2020.

16. The Renewable Energy Hub USA. (2020). *Heat recovery efficiencies*. https://www.renewableenergyhub.us/heat-recovery-systems-information/heat-recovery-system-efficiencies.html. Accessed 12 Sept 2020.
17. World Business Council for Sustainable Development. (2018). *Dispose energy recovery*. Circular Economy Practitioner Guide. https://www.ceguide.org/Strategies-and-examples/Dispose/Energy-recovery. Accessed 12 Sept 2020.
18. United States Environmental Protection Agency. (2020). *Energy recovery from the combustion of Municipal Solid Waste (MSW)*. https://www.epa.gov/smm/energy-recovery-combustion-municipal-solid-waste-msw. Accessed 12 Sept 2020.
19. United States Environmental Protection Agency. (2017). *National overview: Facts and figures on materials, wastes and recycling*. https://www.epa.gov/facts-and-figures-about-materials-waste-and-recycling/national-overview-facts-and-figures-materials. Accessed 12 Sept 2020.
20. United States Department of Energy, Office of Energy Efficiency and Renewable Energy. (2016). *Bioenergy basics*. https://www.energy.gov/eere/bioenergy/bioenergy-basics. Accessed 12 Sept 2020.

Chapter 6
Water Use

6.1 Introduction

As the climate changes, access to water will also change. While droughts limit the availability of water, severe storms cause excess water. This results in poor water quality and makes some geographic locations unsafe. This will not only affect the amount of water a material extractor, processor, or manufacturer can use but also may affect where it locates operations.

There are several questions that pertain to water use across life cycle stages (Figs. 6.1 and 6.2):

- Where is water sourced from, and is there an environmentally preferable source available?
- Can water use be reduced across the life cycle?
- Is water being recovered and returned to the source? If so, does it have the same quality as the original water source?

Just as with energy, water use during a product's life cycle can be banned, limited, or reduced by regulations and certifications. This can be in the form of where the water is drawn from, where the water is discharged, and the total volume of water used. Bans are most often based on narrative, such as "no freshwater will be utilized," whereas limits and reductions may be in terms of narrative, percentages, or specific numeric values.

Water use can also be seen through a more holistic approach: looking at the condition of the ecology at the location of the facility and ensuring a clean water supply for those who live adjacent to where the product is made.

Unlike raw materials or energy, water can be used and returned to a source without reducing the overall volume of water. This is why the quality of a manufacturer's discharged effluent is a key sustainability consideration unique to water use.

H. R. Roth et al., *The Green Building Materials Manual*,
https://doi.org/10.1007/978-3-030-64888-6_6

INITIATIVES AND IMPACTS: WATER USE

Water

Extraction: Water Sourcing
Body of Water Protection
Fresh Water vs. Potable Water
Manufacturing and Use:
Volume Reduction
Water Audits
Water Efficient Tools, Equipment, & Appliances
Net Zero Water Use

Origin: Ecology, Social Accountability
Embodied Water
Water Footprint
Water Stewardship
Worker Water Supply
End of Life:
Water Recycling
Waste Water Quality
Antidegradation Requirements
Water Quality Testing Requirements

Fig. 6.1 Initiatives and impacts for water

WATER USE DURING LIFE CYCLE

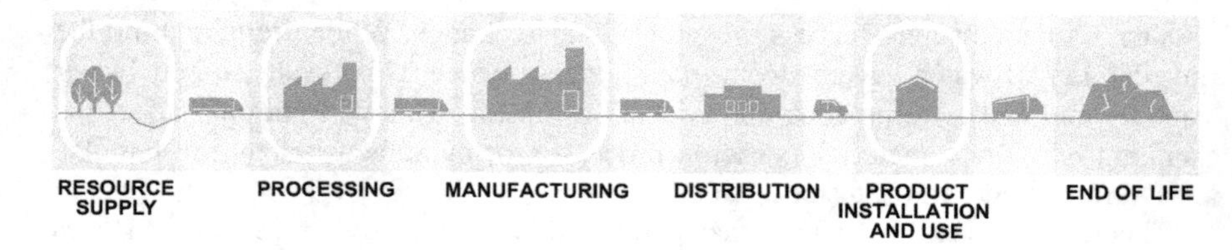

Fig. 6.2 Water use during life cycle stages

The US Clean Water Act

In 1824, a Supreme Court decision ruled that the federal government had the right to control navigation across the United States, including along rivers and across harbors. Legislation, and associated funding, was enacted in 1826 to improve navigation along waterways by dredging and clearing them of obstacles. While most obstacles were natural, some were man-made. In 1899, the Rivers and Harbors Appropriation Act, and the associated Refuse Act, established misdemeanor charges against anyone discharging refuse into navigable waters or their tributaries without a permit. It was the first environmental act involving water in the United States. Fundamental to this was the idea that an entity would be fined if their actions were considered detrimental to another sector's ability to utilize and profit from natural water resources [1].

(continued)

In 1948, the Federal Water Pollution Control Act (FWPCA) was established, being administered by the Department of Interior and authorized by the surgeon general, in cooperation with the Federal Works Administration [2]. The FWPCA worked to develop plans for states and municipalities to control pollution of interstate waters and tributaries. The focus was on limiting the release of raw sewage by constructing new treatment plants. The act's jurisdiction broadened beyond navigation to conserving waters for [3]:

- Public water supplies
- Fish and aquatic life propagation
- Recreation
- Industrial and agricultural uses

This widening of jurisdiction provided protection for these sectors to both utilize and profit from bodies of water. In addition to developing plans for controlling municipal sewage, a permitting system for the discharge of pollution by corporations was established. However, the initial system lacked authority. To achieve enforcement, the federal government pursued lawsuits against corporations using the FWPCA in conjunction with the Refuse Act, which could levy misdemeanor charges against infringement.

In 1972, the FWPCA was reorganized and amended into what is now known as the Clean Water Act (CWA), administered by the Environmental Protection Agency (EPA). The CWA continues to permit the controlled discharge of pollutants by industry, municipality, or other facilities into surface waters [4]:

- Industrial facilities include manufacturing, mining, petroleum extraction, and service industries.
- Municipal facilities include sewage treatment plants and military bases.
- Other facilities include some agricultural uses like feedlots.

These sources of pollution are defined as point sources. Permits regulate the amount and frequency of pollutant discharge. Testing by the point source and disclosure of resulting data to the EPA confirms compliance with the permit. The EPA maintains pollution control by setting wastewater standards and maintaining water quality standards.

Litigation to establish stormwater discharges from cities and industrial entities eventually resulted in the definition of stormwater discharges as point sources in amendments to the act. By contrast, Congress exempted agricultural stormwater and irrigation flows as point sources. This allows agricultural pollutants to enter along the entire length of a waterway without permit.

(continued)

In addition to considering who must hold a permit, the definition of a body of water has also come under question. The CWA holds jurisdiction over interstate bodies of water and their tributaries. In the early 2000s, a series of rulings by the Supreme Court muddied the definition of these water sources in the CWA.

An act like the CWA is legislation voted on by Congress. A regulation interprets and implements a legislative act and is not voted upon by Congress. In 2015, the Obama administration introduced the Clean Water Rule regulation to interpret the CWA by refining the scale and definition of tributaries to include smaller streams, wetlands, and additional sources of water that were seasonally intermittent. This broader definition encompassed 60% of water bodies in the United States and roughly one third of the nation's source for drinking water. Litigation against the Clean Water Rule regulation began immediately. In 2017, the Trump administration announced it would rescind the regulation and in February 2020 repealed the regulation. In turn, litigation in support of the reinstatement of the Clean Water Rule began immediately [5].

6.2 Extraction: Water Sourcing

Water sourcing includes intake and discharge, along with runoff from debris. Water quality has been studied for many decades in the United States. Recent study has also focused on:

- Amount of water usage
- Water usage's effect on bodies of water and water-scarce regions
- Water usage across a global supply chain

Freshwater is the primary source for all human water consumption (Fig. 6.3). This consumption includes drinking water and water that is used in producing food, goods, and materials. Freshwater sources include:

- Rivers
- Streams
- Lakes
- Reservoirs
- Aquifers
- Groundwater

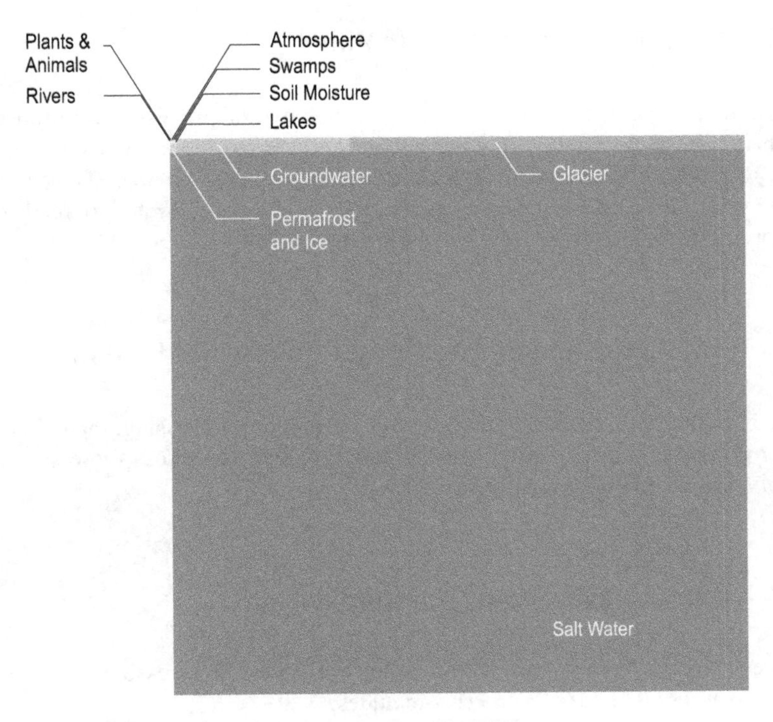

Fig. 6.3 Distribution of types of water by volume on Earth [6]

6.2.1 Body of Water Protection

The EPA provides protections to designated bodies of water. It sets water quality standards based upon the CWA. These water quality standards are the legal basis for controlling pollutants entering into waters, describing the desired condition of the water body, and how that quality is achieved or protected. These characteristics can be defined through numeric benchmarks, or through narrative such as a body of water being free from a negative condition. Bodies of water are protected for:

- Drinking water supply
- Protection and propagation of fish, shellfish, and wildlife
- Recreation
- Agricultural, industrial, navigation, and other designated uses

6.2.2 *Freshwater Versus Potable Water*

Freshwater is water that has been purified through natural processes of filtration and which is accessible directly. Freshwater is not safe for human consumption and must go through additional treatment to be drinkable, or potable. Typically, this means water must be processed by a water treatment plant. Water sourced from a well or aquifer is also usually safe for drinking.

6.3 Manufacturing and Use: Water Volume Reduction

Water is primarily used during the resource extraction, processing, manufacturing, and product use stages of life cycle. Reducing water use across these stages is a primary way to conserve this limited resource.

6.3.1 *Water Audits: Identifying Inefficiencies*

A water audit is synonymous with a water evaluation or assessment. It analyzes water use across a facility [7, 8]. This includes:

- Water intake and discharge – sourcing from and returning to bodies of water or a municipal water system
- Water usage by equipment, appliances, and fixtures

A water audit reveals poorly calibrated or malfunctioning meters and leakage. Additionally, it provides data for equipment, appliances, and fixtures that can be analyzed for efficiency. From collected data, a facility can then develop a strategy to increase efficiency or to mitigate waste.

US water utilities lose large percentages of volume due to aging systems and leakage. In 2004 the EPA estimated that "Average water loss in (utility) systems is 16%, with up to 75% recoverable" [9]. Because so few industries complete water auditing on their premises, it is difficult to know how much water is lost from aging infrastructure or equipment at individual facilities [10, 11].

6.3.2 *Water-Efficient Manufacturing Tools*

Water-efficient tools conserve water. Manufacturing tools and equipment designated as "water-efficient" or "lean" operate with less water by volume [12, 13]. Replacing older tools with water-efficient equipment assessed by the EPA

WaterSense program parallels purchasing energy-efficient equipment that has been evaluated by the EPA Energy Star program [14].

Federal programs are required to purchase water-efficient equipment by executive order. This reduces both potable and non-potable water consumption by government agencies. The Department of Energy's (DOE) Federal Energy Management Program provides specifications for water-efficient product categories for federal agencies. The General Services Administration (GSA) lists these products through their GSA Advantage preferred purchasing website. WaterSense certification fulfills these requirements.

6.3.3 Net Zero Water Use

Net zero is a term borrowed from the energy sector [15]. It sets a limit of returning back to a watershed drainage area the same amount of water that is sourced for processing and manufacturing. This is accomplished through a range of strategies:

- Creating a closed water system, where no water is used from the watershed
- Offsetting the amount of water used with water collected on site – this can be accomplished through:
  - Collecting rainwater
  - Collecting water during manufacturing drying processes, also called dewatering processes
- Using excess water from manufacturing processes on site that would otherwise be discharged. For example, this excess water could be used to irrigate landscaping

Whatever processes are employed, the total amount of water taken from the watershed must equal the total amount returned. A goal is also to preserve the quality of the water, limiting deterioration in a concept similar to retaining the value of waste materials at end of life cycle by a preference of reuse over recycling. While attaining net zero water does not prohibit freshwater use, the goal is to minimize both water sourcing and wastewater discharge into freshwater sources [16].

6.4 Water Origin Related to Ecology and Social Accountability

Beyond pollution of water bodies, sourcing and discharging water can imperil local ecosystems and cultures through the amount of water used. Questions for processors and manufacturers to consider include:

- Is freshwater scarce in the place of operations?
- Would sourcing water at this location threaten the survival of the ecosystem?

- Does running operations at this location put citizens at risk of insecurity for drinking water or food?
- Is the area prone to flooding or drought, and will operations build resiliency or enhance the risk?

6.4.1 Embodied Water

Embodied water or virtual water trade is a concept developed by Tony Allan, a British geographer. It quantifies how much water is required to produce a product or commodity, across the entire life cycle, no matter the location. This evaluation process allows a processor or manufacturer to relate the geographic source of the water and take into account that water is more precious in some environments. Embodied water quantifies water embedded in a product or commodity, giving it value. This value can be considered when a product or commodity is being produced in a water-scarce environment or being imported into a water-scarce environment [17]. For instance, if a ton of cotton is imported from a country, the amount of water needed to produce that cotton, in theory, is also virtually imported. In a dry country, it may make sense to import water-dependent materials and crops, like cotton, rather than attempt to grow them locally.

Embodied water is similar to embodied energy in that it allows a company or consumer to understand the total quantity of water used over a product's life cycle, rather than simply evaluating the water required for operation or use. This allows a manufacturer to also evaluate where large amounts of water are used, identifying opportunities for reducing consumption.

Water use itself may require significant energy consumption. In particular, moving water from a source to a point of use can be energy intensive. This calculation is included in the embodied energy or carbon footprint of a product, not the embodied water or water footprint.

6.4.2 Water Footprint

Water footprint is a concept developed by Arjen Hoekstra in 2002 while working for the UNESCO Institute for Water Health [18]. It is based on the idea of the ecological footprint. Water footprint is a metric that refers to the amount of freshwater used to produce products and services that are consumed by individuals, communities, or businesses. This metric weighs three considerations during calculation:

- The amount of water used
- The amount of water polluted
- The location of water use

By adding location, the water footprint indicates where ecology is impacted across the globe. This location data can be cross-referenced to evaluate whether specific water use for a product is best aligned with the needs of a local community or ecology. Additionally, water footprint provides a quantification for embodied water, a concept similar to embodied energy.

6.4.3 Water Stewardship

Water stewardship was developed by the Alliance for Water Stewardship [19]. The organization states "We define water stewardship as the use of water that is socially and culturally equitable, environmentally sustainable and economically beneficial." The Alliance for Water Stewardship has established a standard by which processors and manufacturers can have their operations evaluated for water. That evaluation includes the effects that water use has on social equity within the community.

Managing Water Risk in Supply Chain

By developing a water resilience plan, processors and manufacturers decrease risk in their supply chain in areas prone to drought or flooding. This is useful for companies dependent upon agriculture or needing water for processing. Most important is understanding the vulnerability of water sources at stages across the supply chain. By reducing risk, moving operations away from areas prone to water scarcity, or by working with governments to protect water sources, resiliency increases across the supply chain. Just as it is important to collect data for ingredients and chemical makeup through all tiers of the supply chain, water data is also important. Unfortunately, because of commodity exchange and multilayered supply chain, this data is rarely tracked [20].

6.4.4 Worker Water Supply

Extraction and manufacturing facilities have a responsibility to monitor and provide adequate drinking water, sanitation, and hygiene to workers. It is important that water supply meets water quality standards provided by the Safe Drinking Water Act both in the United States and extending to processor's and manufacturer's facilities outside of the United States [21].

Growing Cotton in Arizona and Texas

Today a large percentage of cotton in the United States is grown in the arid panhandle and far west of Texas. As it turns out, cotton suffers from heat stress in humidity and thrives in dry air [22, 23]. The issue? In arid regions, this thirsty crop often must be irrigated from groundwater and wells.

The Ogallala aquifer is ancient and enormous, stretching under much of the US plains, including Texas. In 1904, Texas established groundwater rights similar to English water rights: a landowner has the right to pump and use as much water as he or she chooses beneath his or her property, no matter how that usage impacts neighboring landowners or water levels throughout the region [24]. Essentially if a farmer does not use water, someone else will.

- Today, technology provides access to the depths of the aquifer.
- The 1904 law results in no associated cost per volume of water used.
- The weather provides a prime growing environment for cotton.

The result? Water-thirsty crops like cotton are emptying the Ogallala aquifer, and there is no governmental, economic, or legal incentive to halt the process.

This is not the first time governments have supported growing cotton in the desert. In the 1860s as Civil War broke out in the United States, cotton mills in England needed cotton fiber, and the cost for fiber increased substantially. Decades previously in Egypt, a French engineer noticed a few cotton plants in a formal garden. The plants were an unusual cultivar with extremely long fibers. Economically depressed since the fall of the Ottoman Empire, the Egyptian government heard of this discovery and began irrigating land adjacent to the Nile and cultivating the long-fibered cotton. When the Civil War interrupted US cotton production, Egypt took advantage of the situation. Within decades Egyptian cotton amounted to more than 90% of Egypt's export revenues [25, 26].

In 1908, the USDA began a project to directly compete with Egyptian cotton. The agency contracted with the Native American Pima tribe to plant cotton in the Arizona desert. Today Pima cotton is recognized for its quality and softness [27]. What is less known is that the cultivar is heavily supported by the US government through subsidies, payment of insurance, and loan forgiveness to support local farmers [28, 29]. Additionally, irrigation relies upon an aquifer that is being drained by agriculture [30, 31]. "The use it or lose it" rules for water rights, which mirror those of Texas, in conjunction with financial incentives from the government make it too lucrative for farmers to give up Pima cotton as a crop, even if this water-needy plant is being farmed in the desert.

6.5 End of Life: Water Recycling and Water Quality

6.5.1 *Water Recycling*

Municipal treatment systems supply potable water that is suitable for drinking to the community. This includes supplying potable water to processing plants, manufacturing facilities, and office buildings. Often these buildings use potable water for non-potable functions, for example:

- Washing and cleaning raw materials
- Supplying water to air conditioning systems and other equipment
- Watering landscaping
- Flushing toilets

Employing a water recycling system decreases the volume of water needed from a municipal treatment facility [32]. It also uses water for several purposes before it is discharged back into a municipal sewage system. In the case of residential or office use, water from kitchen facilities, baths, and restroom sinks can be collected and reused as gray water. This water can irrigate landscaping or be used in black water functions such as toilets. This decreases treatment requirements and can provide on-site water sourcing. Processing and manufacturing equipment can use water as a source from:

- Reclaimed gray water
- Non-potable water
- Storm water runoff
- Discharge from natural resource extraction such as mining
- Water from dewatering processes – collecting excess water that results from drying materials like wood or fabric fibers

The EPA does not regulate water recycling. Some state and municipal regulations encourage water recycling through tax incentives. Some sustainability standards offer points toward certification when a water recycling program is being used.

6.5.2 *Water Quality and Character*

The measurement of water quality occurs after water is used for extraction, manufacturing, or irrigation, just prior to discharge back into a body of water [33, 34, 35]. Water quality and character include contamination through (Table 6.1):

- Chemicals and radiological matter
- Turbidity (clarity)
- Temperature
- pH (acidity)

Table 6.1 Water contaminants and their effect on bodies of water

Contamination	Action	Effect
Chemicals and Radiological Matter	Chemical and radiological matter introduced into a water source	Impacts human health (see Chap. 8) May fall below EPA benchmarks for toxicity but still degrade water quality
Turbidity (Clarity)	Discharge water speed can cause turbulence which erodes soil Particulates may be released in discharge water Turbidity is the resulting suspension of solids or particulate matter in water	Change the clarity of water. Decreases photosynthesis through water to decrease plant growth Increases opportunity for metals and microbes to attach to particulates, increasing toxicity
Temperature	Water discharge can be at a heightened or lowered temperature	Temperature change affects plant growth Temperature change attracts or discourages aquatic life Heightened temperature decreases oxygen retention
pH (Acidity)	Discharge can include pollutants beneath or unmeasured by EPA toxicity benchmarks but which change water acidity	Change in acidity can increase the solubility of nutrients and heavy metals in water Acidity can lead to water incorporating toxic substances not initially present in the discharge, but present in river or lake bottom soil Acidity can increase or decrease nutrients available to aquatic plants and organisms or affect a plant's ability to metabolize those nutrients
Dissolved Oxygen (Aeration)	Discharge can include pollutants beneath or unmeasured by EPA toxicity benchmarks but which affect dissolved oxygen	Pollutants can decrease dissolved oxygen levels Low oxygen can decrease plant growth or affect organisms When oxygen falls to extremely low levels, called hypoxia, aquatic plants and organisms die, resulting in a "dead zone." See Chap. 7
Excess Organism Growth	Discharge can include pollutants beneath or unmeasured by EPA toxicity benchmarks but which add nutrients from substances like fertilizers	Excess nutrients can lead to algae growth, which causes turbidity Turbidity decreases photosynthesis and affects plant growth Excess algae growth decreases dissolved oxygen levels Low oxygen can decrease plant growth or affect organisms In extreme cases of low oxygen, hypoxia can occur, causing dead zones

- Dissolved oxygen (aeration)
- Organism growth due to excess nutrients

The EPA collects data submitted by facilities for water quality of tap water and pollution point sources along bodies of water. The Food and Drug Administration (FDA) measures water quality of bottled water. The US Geological Survey (USGS) measures water quality for selected surface and groundwater sites [36].

6.5.3 Antidegradation Requirements

Antidegradation means that water quality must be maintained to a level that [37, 38]:

- Previous uses, such as fishing and recreation, can continue
- Certain previously established aquaculture uses such as propagating shellfish can continue
- Resident species, plants, or animals living in the water, or adjacent to the water, cannot have increased mortality, abnormal growth, or changes in reproductive function

An objective of the CWA is to "maintain the chemical, physical and biological integrity of the Nation's waters" [39]. In addition to federal regulation, individual states maintain antidegradation policies.

Bodies of water considered having a better than baseline quality are defined as "high quality." The EPA encourages that states maintain these bodies of water at their current condition, although the EPA allows some degradation in response to projects that encourage economic development. Bodies of water designated as "Outstanding National Resource Waters" receive the highest level of protection under antidegradation requirements. The EPA allows no new or increased discharges from point sources, with the exception of temporary or short-term activities (for instance, constructing a building).

6.5.4 Water Quality Testing Requirements

In the United States, the National Pollutant Discharge Elimination System (NPDES) permit program established by the EPA provides permits to processors and manufacturers who discharge pollutants into a body of water at a point source [40]. When the NPDES permit is written, testing frequency is based upon:

- The amount of discharge
- The variability of the volume of discharge
- The variability of the content of discharge
- The nature of the pollutants
- The history of compliance of a company or entire sector with EPA regulations

- The cost of monitoring and testing
- The design of the capacity of the treatment facilities
- Into what type of body of water the discharge is allowed (high quality, outstanding national resource)

Some processors and manufacturers elect to install a water effluent management system that provides continuous pollution and discharge monitoring. While continuous effluent monitoring is required for drinking water production by regulation, such a system for other water users is not required.

References

1. Franz, A. (2010). Crimes against water: The Rivers and Harbors Act of 1899. *Tulane Environmental Law Journal, 23*(2), 255–278. http://www.jstor.org/stable/43294085. Accessed 12 Sept 2020.
2. United States Fish and Wildlife Service. (1948–1987). *Federal Water Pollution Control Act (Clean Water Act). Digest of Federal Resource Laws of Interest to the U.S.* Fish and Wildlife Service. https://www.fws.gov/laws/lawsdigest/FWATRPO.HTML. Accessed 12 Sept 2020.
3. United States Environmental Protection Agency. (2020). *Water topics*. https://www.epa.gov/environmental-topics/water-topics. Accessed 12 Sept 2020.
4. United States Environmental Protection Agency. (1972). *Summary of the Clean Water Act 33 U.S.C. §1251 et seq*. https://www.epa.gov/laws-regulations/summary-clean-water-act. Accessed 12 Sept 2020.
5. Friedman, L., Davenport, C. (2019). *Trump administration rolls back clean water protections*. The New York Times. https://www.nytimes.com/2019/09/12/climate/trump-administration-rolls-back-clean-water-protections.html. Accessed 12 Sept 2020.
6. Shiklomanov, I. (1993). *Where is earth's water? United States geological survey, from world fresh water sources, water in crisis: A guide to the world's fresh water sources*. New York: Oxford University Press. https://www.usgs.gov/special-topic/water-science-school/science/where-earths-water?qt-science_center_objects=0#qt-science_center_objects. Accessed 12 Sept 2020.
7. Wiant, C. (2017). *Water loss: Challenges, costs, and opportunities*. Water Quality & Health Council. https://waterandhealth.org/safe-drinking-water/water-loss-challenges-costs-opportunities/. Accessed 12 Sept 2020.
8. Office of Water, Office of Ground Water and Drinking Water, Drinking Water Protection Division. (2018). *Drinking water infrastructure needs survey and assessment sixth report to Congress*. United States Environmental Protection Agency. https://www.epa.gov/sites/production/files/2018-10/documents/corrected_sixth_drinking_water_infrastructure_needs_survey_and_assessment.pdf. Accessed 12 Sept 2020.
9. United States Environmental Protection Agency, Office of Water. (2013). *Water audits and water loss control for public water systems*. EPA816-F-13-002. https://www.epa.gov/sites/production/files/2015-04/documents/epa816f13002.pdf. Accessed 12 Sept 2020.
10. Nevada Governor's Office of Energy. (2020). *Public building water auditing best practices*. http://energy.nv.gov/uploadedFiles/energynvgov/content/Programs/Public%20Building%20Water%20Auditing%20Best%20Practices.pdf. Accessed 12 Sept 2020.
11. Alliance for Water Efficiency. (2017). *The water audit*. Utility Water Loss Control. https://www.allianceforwaterefficiency.org/resources/topic/water-audit. Accessed 12 Sept 2020.

12. United States Environmental Protection Agency. (2020). *Lean & water toolkit: Appendix A water efficiency resources and technical assistance providers*. https://www.epa.gov/sustainability/lean-water-toolkit-appendix. Accessed 12 Sept 2020.
13. United States Environmental Protection Agency, Office of Water. (2000). *Using water efficiently: Ideas for industry*. EPA832-F-99-081. https://www.epa.gov/sites/production/files/2017-03/documents/ws-ideas-for-industry.pdf. Accessed 12 Sept 2020.
14. United States Environmental Protection Agency. (2020). *WaterSense*. https://www.epa.gov/watersense. Accessed 12 Sept 2020.
15. United States Environmental Protection Agency. (2020). *Net zero concepts and definitions*. https://www.epa.gov/water-research/net-zero-concepts-and-definitions. Accessed 12 Sept 2020.
16. United States Department of Energy, Office of Energy Efficiency and Renewable Energy. (2020). *Net zero water building strategies*. https://www.energy.gov/eere/femp/net-zero-water-building-strategies. Accessed 12 Sept 2020.
17. Water Footprint Network. (2020). *Virtual water trade*. https://waterfootprint.org/en/water-footprint/national-water-footprint/virtual-water-trade/. Accessed 12 Sept 2020.
18. Water Footprint Network. (2020). https://waterfootprint.org/en/. Accessed 12 Sept 2020.
19. Alliance for Water Stewardship. (2020). *The AWS International Water Stewardship Standard 2.0*. https://a4ws.org/the-aws-standard-2-0/. Accessed 12 Sept 2020.
20. Bateman, A. (2016). *Does your supply chain risk management strategy hold water? MIT sloan management review*. https://sloanreview.mit.edu/article/does-your-supply-chain-risk-management-strategy-hold-water/. Accessed 12 Sept 2020.
21. United States Environmental Protection Agency. (1974). *Summary of the Safe Drinking Water Act 42 U.S.C. §300f et seq*. https://www.epa.gov/laws-regulations/summary-safe-drinking-water-act. Accessed 12 Sept 2020.
22. Elias, E. et al. (2012). Cotton in the Southwestern United States. SW Hub Production Library, United States Department of Agriculture. https://swclimatehub.info/bulletin/cotton-southwestern-us. Accessed 12 Sept 2020.
23. Texas A&M Agrilife Extension. (2020). *Cotton production regions of Texas*. Cotton Insect Management Guide Teaching, Research, Extension and Service. https://cottonbugs.tamu.edu/cotton-production-regions-of-texas/. Accessed 12 Sept 2020.
24. Templar, O. (2020). *Water law*. Handbook of Texas Online, Texas State Historical Association. https://www.tshaonline.org/handbook/entries/water-law. Accessed 12 Sept 2020.
25. The New York Times. (1864). Egyptian cotton: Its modern origin and the importance of supply. *The New York Times, 26*, 1864. https://www.nytimes.com/1864/06/26/archives/egyptian-cotton-its-modern-origin-and-the-importance-of-the-supply.html?login=email&auth=login-email. Accessed 12 Sept 2020.
26. Schwartzstein, P. (2016). How the American civil war built Egypt's vaunted cotton industry and changed the country forever. *Smithsonian Magazine*. https://www.smithsonianmag.com/history/how-american-civil-war-built-egypts-vaunted-cotton-industry-and-changed-country-forever-180959967/. Accessed 12 Sept 2020.
27. Nolte, K. (2007). *Pima cotton*. https://cals.arizona.edu/fps/sites/cals.arizona.edu.fps/files/cotw/Pima_Cotton.pdf. Accessed 12 Sept 2020.
28. Cowling, M. (2020). *Weak monsoon means better cotton*. Tri-Valley Dispatch. https://www.pinalcentral.com/trivalley_dispatch/farm_and_ranch/weak-monsoon-means-better-cotton/article_65f3e9f7-8604-5167-b394-e3d11ad1f037.html. Accessed 12 Sept 2020.
29. Lustgarten, A., Sadasivam, N. (2015) *Holy crop: How federal dollars are financing the water crisis*. https://projects.propublica.org/killing-the-colorado/story/arizona-cotton-drought-crisis. Accessed 12 Sept 2020.
30. Braxton Little, J. (2009). *The Ogallala Aquifer: Saving a Vital U.S. Water Source*. Scientific American, SA Special Editions 19, 1s, 32–39. https://www.scientificamerican.com/article/the-ogallala-aquifer/. Accessed 12 Sept 2020.

31. United States Department of Agriculture, National Resources Conservation Service. (2016). *Ogallala aquifer initiative*. https://www.nrcs.usda.gov/wps/portal/nrcs/detailfull/national/programs/initiatives/?cid=stelprdb1048809. Accessed 12 Sept 2020.
32. United States Environmental Protection Agency. (2020). *Basic information about water reuse*. https://www.epa.gov/waterreuse/basic-information-about-water-reuse#basics. Accessed 12 Sept 2020.
33. United States Geological Survey. (2020). *Turbidity and water*. https://www.usgs.gov/special-topic/water-science-school/science/turbidity-and-water?qt-science_center_objects=0#qt-science_center_objects. Accessed 12 Sept 2020.
34. United States Department of Commerce, National Oceanic and Atmospheric Administration (2019) *Hypoxia*. https://oceanservice.noaa.gov/hazards/hypoxia/
35. United States Geological Survey. (2020). *Water quality information by topic*. https://www.usgs.gov/special-topic/water-science-school/science/water-quality-information-topic?qt-science_center_objects=0#qt-science_center_objects. Accessed 12 Sept 2020.
36. United States Geological Survey. (2020). *USGS water-quality data for the nation*. https://waterdata.usgs.gov/nwis/qw. Accessed 12 Sept 2020.
37. United States Environmental Protection Agency. (2020). *What are water quality standards?* https://www.epa.gov/standards-water-body-health/what-are-water-quality-standards. Accessed 12 Sept 2020.
38. United States Environmental Protection Agency, Office of Water. (2012). *Water quality standards handbook chapter 4: Antidegradation*. EPA-823-B-12-002. https://www.epa.gov/sites/production/files/2014-10/documents/handbook-chapter4.pdf. Accessed 12 Sept 2020.
39. United States Environmental Protection Agency. (2020). *Clean Water Act (CWA) and Federal Facilities*. https://www.epa.gov/enforcement/clean-water-act-cwa-and-federal-facilities. Accessed 12 Sept 2020.
40. United States Environmental Protection Agency (2010) *Chapter 8: Monitoring and reporting conditions*. NPDES Permit Writers' Manual. https://www3.epa.gov/npdes/pubs/pwm_chapt_08.pdf. Accessed 12 Sept 2020.

Chapter 7
Emissions

7.1 Introduction

Emissions refer to the chemicals and pollutants released into the air and water throughout the life cycle of a building material that results in degradation of the environment and negative impacts on human health (Fig. 7.1). Emissions are outputs at every stage in the life cycle, from extraction to end of life (Fig. 7.2).

There are two primary types of emissions in the context of building materials:

1. Product (direct) emissions are a direct result of the ingredients used in a product. Volatile organic compounds (VOCs) are a primary example of this type of emissions. These will be discussed in more detail in Chap. 8.
2. Life cycle (indirect) emissions are a byproduct resulting from the processes of extraction, manufacturing, transportation, or end of life for a product. Greenhouse gas (GHG) emissions, including carbon dioxide, are a primary example of life cycle emissions.

Both types of emissions are difficult to measure. For example, measuring the VOC emissions from a product requires a standardized method of testing, placing a sample in a sealed chamber with controlled airflow, temperature, and humidity for a specific duration of time. This ensures testing conditions are controlled to produce a standardized measurement.

Indirect measurement of emissions that are a byproduct of processing and manufacturing is more difficult. It is not possible to control the environmental conditions surrounding an entire factory for accurate testing. Instead, scientists utilize data collected for a region related to soil quality, water quality, and other characteristics. Scientists relate the collected data to baseline measurements to understand the effect of processing and manufacturing in the area. In some cases, emissions can be controlled at the source. A good example is the release of effluent, also know as liquid waste, when it is discharged from a facility.

H. R. Roth et al., *The Green Building Materials Manual*,
https://doi.org/10.1007/978-3-030-64888-6_7

INITIATIVES AND IMPACTS: EMISSIONS

Fig. 7.1 Initiatives and impacts for emissions

EMISSIONS DURING LIFE CYCLE

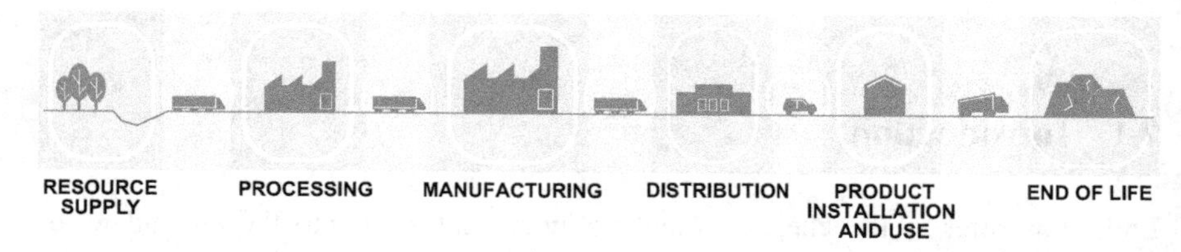

Fig. 7.2 Emissions during the life cycle

The most common method for measuring and tracking reductions of life cycle emissions is an environmental accounting method called life cycle assessment (LCA), explained in further detail in Chap. 3. LCA estimates the emissions resulting from each stage of the life cycle of a product by assigning standardized impact factors to each input and output to characterize and estimate the total impact. These life cycle impacts are typically expressed in equivalent units noted as "eq" or "e" that express the impact of one substance relative to one unit of another substance. Table 7.1 provides an overview of commonly tracked life cycle impacts, their equivalent unit, and the scale at which they impact the ecosystem or humans.

7.2 Carbon Dioxide and Greenhouse Gas Emissions

A report by the United Nations Environment Programme (UNEP) in 2017 identified the building industry as the largest single contributor to global warming. Buildings are responsible for 39% of global carbon emissions, with the building materials industry accounting for 11% of total global emissions [1]. The building industry has a vital role in preventing catastrophic climate change, and it is clear that includes reducing the emissions from the extraction, processing, manufacturing, transportation, and end of life stages for building materials and products. Carbon emissions are a primary example of why defining methods and strategies for tracking and reducing emissions is important despite the technical challenges presented by estimating these emissions throughout complex product supply chains.

Table 7.1 Common life cycle impacts

Life Cycle Impact	Equivalent Unit	Scale of Impact	Human or Ecosystem Damage?
Climate Change/Global Warming Potential	**Kg CO_2eq**	Global	Human Ecosystem
Acidification Potential	**Kg SO_2eq** Moles H+	Regional	Ecosystem
Eutrophication Potential	**Kg N eq** Kg P eq	Local	Human Ecosystem
Stratospheric Ozone Layer Depletion Potential	**Kg CFC-11 eq**	Global	Human
Potential Formation of Tropospheric Ozone/Smog	**Kg ethane eq** Kg NO_x eq Kg O_3 eq	Local	Human

7.2.1 Global Warming Potential

Global warming refers to the increased global average temperature over time as a result of trapped heat (i.e., infrared radiation) by GHG in the atmosphere. There are over a hundred GHG, each with a different chemical composition. Global warming caused by human activity, such as the manufacture of building materials, is referred to as anthropogenic global warming. The phrase climate change more broadly describes the changes resulting from global warming. Global warming contributes to severe heat waves, floods, droughts, and other weather patterns resulting from changing temperatures.

The primary anthropogenic sources of GHG are fossil fuel combustion and land use change, such as deforestation. As discussed above, the building industry has a large role in these sources, as much of the fossil fuel combustion is related to constructing and operating buildings. Carbon dioxide (CO_2) is the most well-known GHG, but methane (CH_4), nitrous oxide (N_2O), and even water vapor (H_2O) are also GHG that contribute to global warming. The EPA maintains a list of GHG, coordinated with the framework established by the United Nations Framework Convention on Climate Change (UNFCCC) [2]. These substances are known to trap and absorb heat, resulting in global warming.

All GHG can also be described by their global warming potential (GWP). The GWP of a product provides a way to compare the possible global warming impacts of different GHG. Different GHG can absorb different amounts of solar radiation, and they can hold that energy for different periods of time. The GWP relates the amount of heat the emissions of 1 ton of a GHG will contribute compared to the amount of heat 1 ton of carbon dioxide (CO_2) contributes. The larger the GWP, the more the specific GHG causes global warming potential compared to CO_2. The timescale is set at 100 years, though many GHG persist for a much shorter period of time. GWP provides a common unit of measure across GHG, allowing easier comparison [3].

Manufacturing contributes four primary greenhouse gases: carbon dioxide, methane, nitrous oxide, and fluorinated gases (Table 7.2) [4, 5].

7.2.2 Embodied Carbon

A product's embodied carbon refers to the sum of emissions released across a product's life cycle. This can also be described as a product's carbon footprint, climate impact, or global warming potential. A product's embodied carbon can quantify a portion of its life cycle, such as from cradle to gate (Stages A1–A3; see Chap. 3), or it can refer to the sum of emissions from "cradle to grave" across its entire life cycle. Strategies to reduce the embodied carbon of building materials include:

- Reusing existing buildings and materials
- Reducing the total quantity of materials used through design strategies aimed at eliminating excess materials and using spaces more efficiently

Table 7.2 Predominant greenhouse gases (GHG)

Greenhouse Gas (GHG)	Source and Effect
81% Carbon Dioxide (CO_2) GWP 1/100 Years	Burning of fossil fuels like coal, natural gas, and oil
	Industrial processes with chemical reactions that produce CO_2, including cement manufacture
	Iron and steel represent about 5% of total global emissions due to burning fossil fuels – this includes architectural products and infrastructure
	Cement accounts for about 8% of total global emissions due to fossil fuel combustion and cement's chemical reactions – this includes architectural products and infrastructure
	Plant decay produces CO_2
	Plant growth and other biological processes absorb CO_2, removing it from the atmosphere. This process is referred to as carbon sequestration
10% Methane (CH_4) GWP 28 to 36/100 Years	Production and transport of coal, natural gas, and oil
	Digestion processes by livestock and other agricultural practices
	Decay of organic waste in landfills
7% Nitrous Oxide (N_2O) GWP 265–298/100 Years	Agricultural and industrial activities
	Combustion of fossil fuels and solid waste
	Wastewater treatment
3% Fluorinated Gases GWP 1000 to 10,000 s/100 Years	Entirely synthetic (i.e., man-made) gases emitted during industrial processes
	Hydrofluorocarbons (HFCs), perfluorocarbons (PFCs), sulfur hexafluoride, and nitrogen trifluoride
	HFCs are used as a replacement for ozone-depleting substances (see Sect. 7.5) that were banned by the Montreal Protocol. This results in an unfortunate tradeoff between ozone depletion and global warming

- Reducing the amount of high-impact materials used
- Optimizing material use through selecting low-carbon or carbon sequestering materials

The identification of high-impact materials or low-carbon materials can be challenging, as impacts and opportunities are dependent on the building's program, design, climate and geographic location, and other building characteristics. The largest contributor to a building's embodied carbon is typically the structure, followed by the envelope and interior finishes (Table 7.3).

A consensus around which materials are low-carbon or carbon sequestering is still developing, as research is relatively recent. The Carbon Smart Materials Palette developed by Architecture 2030 identifies examples of low-carbon materials such as [6]:

- Bamboo, including engineered structural components, finishes, acoustic panels, and structural panels
- Hempcrete wall systems
- Sheep's wool insulation
- Straw-bale insulation
- Wood (if sustainably harvested), including applications for structure, envelope, and finishes

As heating, cooling, and other building operations become increasingly efficient, embodied carbon in materials and products make up a larger and larger fraction of emissions produced within the construction sector. For the estimated 2 trillion square feet of new construction that will occur between 2020 and 2050, embodied carbon will be responsible for 49% of the total carbon footprint [7]. One key difference between embodied carbon emissions and operational carbon emissions from energy use in buildings is that the operations of a building can become more efficient over time, reducing the total operational carbon. It is impossible to reduce embodied carbon over the life of a building, as the emissions have already occurred by the time a building material is used for construction.

Table 7.3 Typical hot spots

Building Component	Typical Hot Spots
Building Structure	Concrete Steel Wood (except from sustainably harvested forests)
Building Envelope	Insulation Glass and aluminum cladding systems
Interior Finishes	Carpet Gypsum board

7.2.3 *Greenhouse Gas Accounting and Emissions Reporting*

Emissions reporting can be specific to a processor or manufacturer, or it can also include the data from suppliers. It can consider just the manufacturing stage of life cycle or emissions across an entire life cycle assessment, including emissions produced while transporting a product from one stage to the next.

GHG emissions are grouped into three categories or "scopes" that are defined by the Greenhouse Gas (GHG) Protocol for the purpose of emissions reporting. The GHG Protocol provides a consistent standard for collecting and measuring emissions data (Table 7.4). It was established by the World Business Council for Sustainable Development and the World Resources Institute.

Scope 1, 2, and 3 emissions vary widely depending on an organization's sector, size, and geographic location. Scope 1 and 2 emissions are more commonly reported than Scope 3 because Scope 3 has recently been adopted. [8].

7.2.4 *Carbon Offsets*

After measuring the carbon emissions attributed to a product's life cycle, a company may elect to purchase carbon offsets. Carbon offsets are a way to balance carbon emissions by funding projects that reduce GHG or prevent projects that would otherwise create GHG. Carbon offsets are measured in metric tons of carbon dioxide equivalent (Mt CO2e). 1 Mt CO2e is equal to 1000 kgCO_2eq (see Chap. 3, Life Cycle Assessment, to learn more about equivalency and the calculation of kgCO_2eq).

Purchases of carbon offsets can help toward certification to a standard. These purchases can be in lieu of meeting an emissions limit or ban. In some cases, companies make a voluntary purchase to demonstrate their support of decreasing emissions and meeting public goals.

Table 7.4 GHG Protocol "scope" emissions

GHG Protocol Scope	Definition
Scope 1 Emissions	Direct emissions Owned or controlled source, such as natural gas or diesel fuel
Scope 2 Emissions	Indirect emissions Purchased electricity, steam, heating, and cooling consumed by the reporting company
Scope 3 Emissions	All other indirect emissions Along a company's value chain Emissions from waste, employee commuting, air travel, and purchased goods

A number of organizations have created standards to verify that projects represented by the carbon offsets are valid. Several of the available standards include:

- The Verified Carbon Standard
- The Gold Standard
- The American Carbon Registry
- The Clean Development Mechanism (CDM)

The CDM was established by the Kyoto Protocol and verifies the authenticity of offsets by providing CDM Certified Emissions Reductions certificates. A key component of many standards is a registry ensuring that each carbon offset is only purchased once. This prevents multiple companies from purchasing credits for the same project, which would result in a value lower than the tons of CO_2 attributed to the project [9].

Kyoto Protocol and Paris Agreement

The Kyoto Protocol to limit emissions was adopted in 1997 and became enforced in 2007. The Protocol set upper limits for the production of GHG for countries that signed the protocol. Different countries had different limits depending upon their level of development and ability to fight global warming.

The European Union Emissions Trading Scheme was established in 2005. It provided a cap and trade system, where over 11,000 individual emissions generating factories, power stations, and other facilities agreed to a total emission limit. If that limit was not met, then the facilities were required to purchase carbon offsets. When countries who signed the Kyoto Protocol did not meet the established limits, they also were required to purchase offsets [10].

While narratives state the Protocol has been a success, no individual source shows a quantitative reduction of GHG that meets the specific numerical goal of 5% reduction across signatories. A second commitment period, called the Doha Amendment, has been signed by fewer countries. While this commitment period lasts until 2020, the UN has shifted focus to the Paris Agreement [11].

The Paris Agreement's goal is to keep global temperature increase below 2 degrees centigrade from pre-industrial levels. Unlike the Kyoto Protocol, specific percentage decreases in GHG by individual countries are not required. Instead the Paris Agreement requires that individual countries set goals and then develop plans to implement those goals. All 196 signatories of the UNFCCC signed onto the Paris Agreement in 2015, including the United States. President Trump has announced his plan to withdraw from the agreement in November 2020. Since 2017 US regulations have been relaxed or established to be directly contrary to the goals of the Paris Agreement [12].

7.3 Acidification

Acidification is the process of increasing the acidity or pH balance of an environment [13, 14]. This occurs through the direct release of acids (such as nitric acid or sulfuric acid) into an environment or through the release of certain chemicals into the air or water that results in a chemical reaction increasing the acidity of soil or water. The chemicals and reactions most often associated with acidification are:

- Sulfur oxides (SO_x)
- Nitrous oxides (NO_x)
- Ammonia (NH_3)
- Carbon dioxide (CO_2)

A small portion of SO_2 and NO_x are emitted through natural processes, such as volcanoes. However, the major source for SO_2 and NO_x emissions includes:

- Burning of fossil fuels
- Using vehicles and heavy equipment
- Industrial manufacturing
- Electric power generation – which accounts for two thirds of SO_2 and one fourth of NO_X in the atmosphere

Ocean acidification and acid rain are acidification processes that directly affect ecology [15]. Acid rain also affects the built environment. The impact of acidification varies in scale. Ocean acidification is a global phenomenon, whereas acid rain is typically confined to a region within several hundred miles down-wind (or downstream) of the emissions source. The scale of impact depends on regional weather patterns and the preexisting acidity of the soil or water in a particular region (Table 7.5).

Table 7.5 Ocean acidification and acid rain

Acidification Type	Ocean Acidification	Acid Rain
Reaction	CO_2 dissolved in water	SO_2 and NO_X reacting with water and other chemicals in the atmosphere
Acid Produced	Carbonic acid (H_2CO_3)	Sulfuric acid (H_2SO_4) and nitric acid (HNO_3)
Result	Removes calcium (Ca) from an aquatic environment Damages the ecosystem Erodes calcium in coral, shellfish, and other populations whose skeletons are derived from calcium	When acid rain falls, it acidifies the soil Leaches nutrients Limits the ability for many plants and other organisms to thrive Acid rain erodes the surfaces of stone and masonry building materials Discolors architectural metalwork

Britain's Houses of Parliament

To many, the term "London Fog" elicits a romantic image of a misty and mysterious atmosphere enveloping London in the late nineteenth century. In the second half of the nineteenth century, London Fog was made up of water vapor but also pollution from coal combustion. Its pea soup character was caused by the presence of nitrogen oxide, sulfuric oxide, ozone, smoke, and airborne particulates. In fact, the word smog is a contraction of smoke and fog, a word developed to specifically describe London's polluted atmosphere [17]. From 1899 to 1901, Claude Monet completed a series of paintings of the British Houses of Parliament [16]. Typically, Monet's paintings were extremely colorful, and these were no different – the sky was portrayed in reds and yellows. Scientists now assert that these paintings most likely depicted how the sky actually appeared as the sun pierced the London Fog.

Victorian England was a global leader. When London's Houses of Parliament were conceived, it was important that they signified Britain's permanence and importance. The Houses of Parliament were built of limestone. Construction began in the 1840s and completed in 1870. During construction the stone began to erode. The acidity of the London Fog was attacking the building's surfaces. Parliament became concerned that Britain's symbol of democracy was being degraded by pollutants coming from the very advances in technology and industry that were making Britain a world leader. While many complaints had been made about the effects of London Fog on human health, Timothy Hyde, an architectural historian at MIT, writes that it was the erosion of the Houses of Parliament that motivated the government to take action [18]. In 1853, Parliament passed the Metropolitan Smoke Abatement Act, limiting the production of smoke in London. In 1875, Parliament enacted the Public Health Act which included stipulations on limiting smoke across all of Britain. These laws acted as precursors to modern environmental laws that later addressed emissions.

7.4 Nutrient Pollution and Eutrophication

The nutrients nitrogen (N) and phosphorus (P) are keys to healthy plant growth and survival. When excess nutrients enter an environment, they can create accelerated excess growth of algae, weeds, and other biological activity [13, 14]. The major sources of excess N and P are fertilizers, wastewater treatment plants, and runoff from processing plants for food, paper, and other plant pulp. Excess nitrogen is particularly damaging to coastal environments, whereas excess phosphorus is more damaging to freshwater lakes and streams [19].

When nutrient pollution accelerates plant growth in water, often in the form of algae blooms, those plants consume excess amounts of oxygen in the water. This can result in eutrophication and hypoxia (Table 7.6).

Table 7.6 Eutrophication, algae blooms, and hypoxia

Step	Event Due to Excess Nutrients	Explanation
1	Algae Blooms	Excessive growth of algae, often as a result of nutrient pollution, can release airborne and waterborne byproducts toxic to humans
2	Eutrophication	Reduced oxygen in an aquatic environment due to excessive algae or other plant growth
3	Hypoxia	Decreased oxygen to the level that kills plants and animals in an aquatic environment, leading to a dead zone, as a result of Step 2

Table 7.7 Ozone-depleting substances banned by the EPA

Substance	Use	Banned by EPA
Chlorofluorocarbons (CFCs)	Refrigerants Foam blowing agents Solvents	1996
Halons	Fire extinguishing agents	1994

The impacts of eutrophication are typically local, although excess nutrients can impact the stream, river, or lake adjacent to the runoff or travel within a body of water for long distances, impacting downstream and coastal ecosystems.

7.5 (Stratospheric) Ozone Depletion

The stratosphere is the second band of atmosphere out from the surface of the Earth [20]. The stratosphere holds the ozone layer which protects the Earth from the Sun's ultraviolet radiation. Ultraviolet radiation causes negative health impacts such as skin cancer and cataracts.

Ozone depletion occurs when substances released into the atmosphere have a chemical reaction in the stratosphere that produces chlorine [13, 14]. Chlorine destroys ozone (O_3) molecules in the ozone layer of the stratosphere. The primary substances linked to stratospheric ozone depletion include chlorofluorocarbons (CFCs) and halons (Table 7.7). A list of ozone-depleting substances is maintained by the World Meteorological Organization (WMO) and the EPA [21].

Many banned ozone-depleting substances have been replaced by fluorinated gases such as hydrofluorocarbons (HFCs). While these synthetic gases have reduced ozone depletion, fluorinated gases are potent GHG with extremely high global warming potential. Some HFCs have a GWP as high as 12,000 times greater impact than an equivalent amount of CO_2. Few solutions have been found that behave in the same way as banned ozone-depleting substances, but which do not have high GWP.

Similar to global warming, stratospheric ozone depletion is a global issue with global impacts. Ozone-depleting substances released into the atmosphere from any location will accumulate and react in the stratosphere to deplete ozone.

Montreal Protocol
The Montreal Protocol is an international treaty designed to gradually eliminate ozone-depleting substances to prevent stratospheric ozone depletion [22]. Agreed upon in 1987, it became enforced in 1989. The protocol was signed by 197 countries, making it the first treaty in the history of the United Nations to achieve universal ratification. The United States signed the Montreal Protocol in 1987.

Universal adoption and implementation of the treaty was extremely successful and has resulted in a pronounced decrease in ozone-depleting substances. By 2010, 98% of the ozone-depleting substances had been phased out of production. As a result of the limits on CFC production put in place by the Montreal Protocol, recovery of the ozone layer is expected by approximately 2067.

7.6 Photochemical Ozone Creation (Smog) Potential

Stratospheric ozone (O_3) protects human health by preventing radiation. In contrast, ground-level ozone, commonly referred to as smog, negatively impacts human respiratory health, ranging from asthma and respiratory irritation to permanent lung damage.

Ground-level ozone is created through a chemical reaction. Nitrous oxides (NOx), sulfur oxides (SOx), and chemicals with volatile organic compounds (VOCs) react in the presence of heat and sunlight. The primary sources of ground-level ozone are nitrous oxides (NO_x) emitted into the air by cars, power plants, and industrial facilities [13, 14].

Smog creation is a local phenomenon. While industrial emissions must be present, smog also depends on the weather patterns and topography of a region. For example, wind and rain can dissipate smog by spreading pollution or causing it to drop out of the atmosphere, whereas mountains can trap emissions within a smaller region, increasing the likelihood of smog formation.

The Clean Air Act

The Clean Air Act is a law that provides federal agencies with the legal authority to regulate air pollution [23]. During the last 150 years, the understanding of air pollution has changed sharply in the United States [24, 25].

The Industrial Revolution brought the widespread use of coal as fuel. Early air pollution law focused upon smoke produced largely from coal. Coal pollution was easy to see, ranging from black to gray in color, with the darker and thicker smoke determined as a nuisance. Initial efforts to decrease the thickness and darkness of smoke resulted in legislation that required:

- More efficient and better quality coal sources
- Better designed boilers – first based upon very specific design methods and later on performance specifications
- The addition of cleaning and capturing mechanisms – limiting the amount of smoke and particulates released into the air

With greater industrialization, coal use increased in metropolitan areas. The geography around some cities resulted in atmospheric inversion, trapping pollution. A series of events in the middle of the twentieth century led to the focus on air pollution from individual offenders to a more widespread concern:

- 1928 – The US Public Health Service began reporting air pollution levels, stating sunlight was reduced by 20 to 50% on some days depending on the weather.
- 1939 – St. Louis had 9 days of zero visibility due to pollution. This led St. Louis to implement the first laws to limit the use of low-quality coal.
- 1940s – Smog events became commonplace in Los Angeles.
- 1948 – Donora, Pennsylvania, suffered an air pollution event that killed 20 and sickened 40% of the town's 14,000 residents.
- 1952 – The London Fog/Smog event for 4 days in December killed approximately 4,000 and sickened another 100,000. This led to the 1956 British Clean Air Act.
- 1953, 1963, 1965 – New York City suffered pollution events that killed 260, 405, and 168 people, respectively. Many thousands were sickened.

While it was understood that air pollution affected human health, it took large-scale sickening and death for the framework of legislation to move away from nuisance law. The focus shifted from punishing the worst offenders to lessening pollution across all industries. Municipalities and states began enacting legislation to limit air pollution, but many of these local governmental agencies did not have the funding to develop ways to test pollution or enforce laws.

In 1955, Congress enacted the first federal air pollution law, the Air Pollution Control Act. This law did not seek to limit pollution. Instead, it provided a framework and financing strategy to establish research on the scope of

(continued)

air pollution and the establishment of standardized testing methods. The law provided 5 million dollars per year (equivalent to $50 million/year in 2020) in grants to states' academic institutions and staff training programs. Additionally, it provided technical assistance from the federal government. Linking grants to local research garnered widespread state support for this federal legislation.

Research increasingly studied not only visible air pollution but also invisible pollution. Studies included health effects from a range of chemical emissions, establishing benchmarks for dangerous levels. Evidence-based research methods developed protocols for testing emissions.

In 1963, the Clean Air Act authorized research on monitoring and began controlling pollution. A 1967 extension of the law provided the federal government with the authority to conduct ambient monitoring and source inspections of stationary operations and facilities. It focused upon studies of which pollutants specific chemicals emitted and began setting nationwide emission standards. The 1967 Act initiated debate across industry as to whether it was appropriate to set a singular approach to regulating air pollutants or a more nuanced and tailored approach regulating individual industries.

The 1970 Clean Air Act shifted the federal government's role in controlling air pollution. After years of research, the government now issued programs and standards that all states were required to follow:

- The National Ambient Air Quality Standards (NAAQS) – for high-volume air pollutants, such as carbon monoxide, lead, nitrogen dioxide, ozone, particulates, and sulfur dioxide
- State Implementation Plans (SIPs) – methods for states to bring local regions into compliance with NAAQS
- New Source Performance Standards (NSPS) – regulating emissions from new facilities
- National Emission Standards for Hazardous Air Pollutants (NESHAP) – standards for air pollutants other than NAAQS that may cause cancer, reproductive changes, birth defects, or negative environmental conditions

At the same time, President Nixon issued an executive order establishing the EPA. This established a federal agency with authority from the Clean Air Act to interpret the law through regulation and to establish enforcement methods. Initially, the EPA set out to regulate chemicals with high hazard risks to human health. Due to litigation, the EPA was only successful in regulating seven chemicals. Federal funding approved by the law also supported enforcement of state and municipal regulations that interpreted the Clean Air Act. Additionally, the act set new standards for vehicle emissions. This included requiring 1975 vehicles to have carbon monoxide (CO) emissions levels reduced by 90% compared to 1970 vehicles. A 1977 amendment strengthened the act's ability to enforce compliance, especially across regions that were

(continued)

still emitting high-volume air pollutants. The last amendment to the Clean Air Act was in 1990. The amendment continued to support technology- and evidence-based restrictions. It also increased federal authority and responsibility for regulating air pollution. New initiatives included:

- New program to control acid rain
- New program to protect stratospheric ozone by phasing out chemicals that deplete ozone (established directly after the United States signed the Montreal Protocol)
- Providing a structure to issue operating permits for facilities that emit air pollution
- NAAQS strengthened to include NOx and VOCs
- NESHAP extended to 189 chemicals or groups of chemicals termed hazardous air pollutants (HAPs)

References

1. Abergel, T., Dean, B., Dulac, J.(2017). *Global Status Report. UN Environment Programme, the International Energy Agency (IEA) for the Global Alliance for Buildings and Construction (GABC)*. https://www.worldgbc.org/sites/default/files/UNEP%20188_GABC_en%20%28web%29.pdf. Accessed 12 Sept 2020.
2. United Nations Climate Change Committee. (2020). *United Nations framework convention on climate change*. https://unfccc.int/. Accessed 12 Sept 2020.
3. United States Environmental Protection Agency. (2020). *Understanding global warming potentials*. https://www.epa.gov/ghgemissions/understanding-global-warming-potentials. Accessed 12 Sept 2020.
4. United States Environmental Protection Agency. (2020). *Greenhouse gases*. Report on the Environment. https://www.epa.gov/report-environment/greenhouse-gases. Accessed 12 Sept 2020.
5. United States Environmental Protection Agency. (2020). *Overview of greenhouse gases*. Greenhouse Gas Emissions. https://www.epa.gov/ghgemissions/overview-greenhouse-gases. Accessed 12 Sept 2020.
6. Architecture 2030. (2020). *Carbon smart materials palette*. https://materialspalette.org/. Accessed 12 Sept 2020.
7. Architecture 2030. (2020). *New buildings embodied carbon*. https://architecture2030.org/new-buildings-embodied/. Accessed 12 Sept 2020.
8. World Resources Institute. (2020). *Greenhouse gas protocol*. https://ghgprotocol.org/. Accessed 12 Sept 2020.
9. Forest Trends' Ecosystem Marketplace. (2018). *Voluntary carbon markets insights: 2018 outlook and first-quarter trends*. https://www.forest-trends.org/wp-content/uploads/2019/04/VCM-Q1-Report-Final.pdf. Accessed 12 Sept 2020.
10. Johnston, E. (2017). 20 years after Kyoto Protocol, where does world stand on climate? *The Japan Times*. https://www.japantimes.co.jp/news/2017/12/04/reference/20-years-kyoto-protocol-world-stand-climate/#.XmLKi0p7lPY. Accessed 12 Sept 2020.

11. United Nations Climate Change Committee. (2020). *The Paris agreement.* https://unfccc.int/process-and-meetings/the-paris-agreement/the-paris-agreement. Accessed 12 Sept 2020.
12. Gross, S. (2020). *What is the Trump administration's track record on the environment?* Policy 2020 Brookings, Brookings Institution. https://www.brookings.edu/policy2020/votervital/what-is-the-trump-administrations-track-record-on-the-environment/. Accessed 12 Sept 2020.
13. Simonen, K. (2014). *Life cycle assessment.* New York: Pocket Architecture: Technical Design Series, Routledge, Taylor & Francis Group.
14. United States Environmental Protection Agency. (2012). *Tool for the Reduction and Assessment of Chemical and Other Environmental Impacts (TRACI) TRACI version 2.1 User Guide.* EPA/600/R-12/554. https://nepis.epa.gov/Adobe/PDF/P100HN53.pdf. Accessed 12 Sept 2020.
15. United States Environmental Protection Agency. (2020). *What is acid rain?* https://www.epa.gov/acidrain/what-acid-rain. Accessed 12 Sept 2020.
16. Merali, Z. (2006). Monet's art may reveal Victorian London's smog. *New Scientist.* https://www.newscientist.com/article/dn9699-monets-art-may-reveal-victorian-londons-smog/. Accessed 12 Sept 2020.
17. Online Etymology Dictionary (1905) Smog (n.). https://www.etymonline.com/word/smog
18. Hyde, T. (2017). How the 19th-century rebuilding of Britain's Houses of Parliament made air pollution visible. *The Conversation.* https://theconversation.com/how-the-19th-century-rebuilding-of-britains-houses-of-parliament-made-air-pollution-visible-71608. Accessed 12 Sept 2020.
19. United States Environmental Protection Agency. (2020). Nutrient pollution, *The Issue.* https://www.epa.gov/nutrientpollution/issue. Accessed 12 Sept 2020.
20. United States Environmental Protection Agency. (2020). *Basic ozone layer science.* https://www.epa.gov/ozone-layer-protection/basic-ozone-layer-science. Accessed 12 Sept 2020.
21. The Intergovernmental Panel on Climate Change. (2014). *Global warming potential values.* Greenhouse Gas Protocol, IPCC Fifth Assessment Report, 2014. https://www.ghgprotocol.org/sites/default/files/ghgp/Global-Warming-Potential-Values%20%28Feb%2016%202016%29_1.pdf. Accessed 12 Sept 2020.
22. United States Environmental Protection Agency. (2020). *International actions – The Montreal protocol on substances that deplete the ozone layer.* https://www.epa.gov/ozone-layer-protection/international-actions-montreal-protocol-substances-deplete-ozone-layer. Accessed 12 Sept 2020.
23. United States Environmental Protection Agency. (2020). *Evolution of the Clean Air Act.* https://www.epa.gov/clean-air-act-overview/evolution-clean-air-act. Accessed 12 Sept 2020.
24. Center for Chemical Process Safety. (2006). *Historical perspective on air pollution control. Appendix I, safe design and operation of process vents and emission control systems by center for chemical process safety*, John Wiley & Sons, Inc. https://onlinelibrary.wiley.com/doi/pdf/10.1002/0470038071.app9. Accessed 12 Sept 2020.
25. Stern, A. (1982). History of air pollution legislation in the United States. *Journal of the Air Pollution Control Association, 32*(1), 44–61. https://doi.org/10.1080/00022470.1982.10465369. https://www.tandfonline.com/doi/abs/10.1080/00022470.1982.10465369.

Chapter 8
Toxicity and Human Health

8.1 Introduction

Many building materials and products have chemical ingredients. Chemicals can be:

- The primary ingredient, for example, polyester or vinyl in textiles or carpets
- A secondary ingredient such as glue
- A treatment or coating that improves performance such as stain prevention or waterproofing

Many chemicals are toxic substances that must be identified and controlled to prevent health impacts on factory workers, construction workers, and building occupants (Fig. 8.1). To protect people and ecosystems, laws and standards limit, ban, or monitor the use of toxic chemicals and the release of their waste. A toxic chemical or substance may be controlled by:

- Laws and Regulations: A small number of substances are either banned or limited in the United States. Laws and regulations are discussed further in Chap. 10.
- Standards and Certifications: Many voluntary standards and certifications include bans or limits of certain substances, as well as third-party product testing to confirm their removal. Standards and certifications are discussed further in Chap. 10.
- Disclosure: Rather than certify to a standard, some manufacturers publish information about the chemicals in their products. This provides the opportunity to change manufacturing processes to voluntarily remove, limit, or replace toxic chemicals. Transparency and disclosure are discussed further in Chap. 11.

While laws and regulations may regulate what is deemed toxic, the term non-toxic is unregulated. No governmental agency holds the term accountable.

Chemicals can be toxic at different life cycle stages (Fig. 8.2):

- Material Supply or Feedstock: The chemicals or reactants used in chemical formulation to make a material through either synthesis or decomposition.

H. R. Roth et al., *The Green Building Materials Manual*,
https://doi.org/10.1007/978-3-030-64888-6_8

INITIATIVES AND IMPACTS: TOXICITY

Toxicity and Human Health

Hazard versus Risk
Exposure Pathways
Levels of Toxic Impacts
Effects of Toxins
Exposure Limits

Evaluation and Control:
Bans on Specific Chemicals
Chemical Red Lists
Six Chemical Classes
Characterizing, Optimizing, and Managing Chemicals

Fig. 8.1 Initiatives and impacts for toxicity

TOXICITY DURING LIFE CYCLE

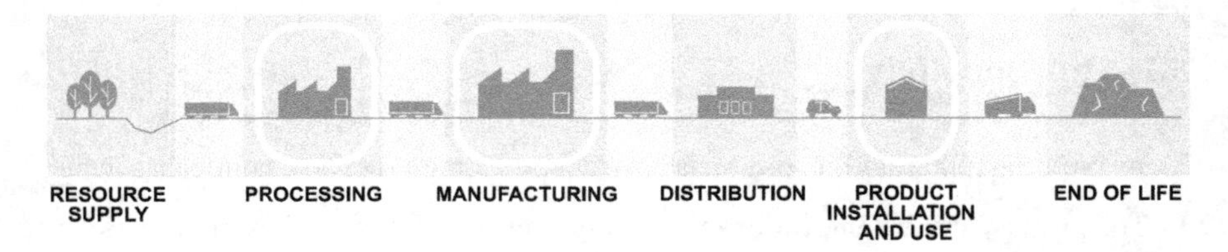

Fig. 8.2 Currently environmental initiatives to limit toxicity focus on the processing, manufacturing, and product use stages of the life cycle

- During Formulation, Post-processing, and Installation: If making a material involves a multistep process, toxic chemicals may occur during formulation but may resolve into inert chemicals or materials in a final product. During the formulation process, one expects that reactions are contained. However, if uncontained, and workers are exposed, the Occupational Safety and Health Administration (OSHA) acts as the regulatory authority to address the exposure. Additional toxicity may be released during post-manufacture processing, such as through sanding, or during installation through field modification of a material. At those stages, the OSHA and Safety Data Sheets (SDS) provide regulatory oversight and direction.
- Products: This includes the chemicals that exist within a final product. These may be inert, or they may deliver toxic exposure.
- Emissions: The airborne or waterborne emissions resulting from a reaction or formulation which are released into the environment.

Current strategies to control the toxicity of building products focus on eliminating or reducing the amount of toxic substances used during processing and manufacture and the release of toxic chemicals during product use once it is installed in a building. This includes initial ingredients (material supply or feedstock), formulation, post-processing, and installation. Currently, there is not a strong focus on regulating or controlling toxicity during the extraction of raw materials or during the demolition of a building, with the exception of asbestos remediation.

8.2 Toxicity Exposure: Hazard Versus Risk

To understand how dangerous exposure to a chemical may be, it is important to differentiate between hazard and risk [1]. A hazard is something that can cause damage or harm under certain conditions. A risk is a chance or probability of being harmed or damaged when exposed to a hazard. Often there is a numerical value associated with a risk based on adjacency or time period of exposure. To reduce a risk, measures to control the hazard can be taken. For instance, a chemical may be toxic during the production of a material. During that time, any hazard of exposure must be controlled to reduce risk to factory workers. This same chemical might be inert, or stable after a product is manufactured. The hazard that the material presented earlier now creates no risk to users in the workplace or at home.

Understanding this allows a consumer to avoid false statements – especially when an organization is claiming a hazard which in reality poses no risk.

8.3 Exposure Pathways

There are three primary ways that humans are exposed to toxic substances:

1. **Ingestion:** Eating or drinking a toxic substance, or a liquid or food that is contaminated with a toxin. Toxic substances can reside in the fatty tissue, internal organs, or bloodstream after ingestion.
2. **Inhalation:** Breathing in toxins in the form of emissions, vapors, particles, or mists. Once inhaled, toxic substances can affect mucous membranes and lungs or enter the bloodstream.
3. **Dermal Exposure:** Toxins can contact the skin or come into contact with the eyes, particularly when found in ubiquitous household objects and building materials. Dermal exposure can result in irritation or burns.

Exposure Through Ingestion and Contact: Persistent Bioaccumulative Toxins (PBTs)

Persistent bioaccumulative toxic substances (PBTs) are toxic substances that are controlled by government agencies. PBTs collect in plants and animals. Exposure occurs through ingestion or dermal exposure.

In animals and people, PBTs accumulate in fatty tissues and do not metabolize at a normal rate. Their concentration often increases up the food chain. Marine mammals, fish, bottom dwellers, and certain birds are easily affected. PBTs cause cancer, genetic changes, nervous system disorders, reproductive issues, and developmental issues.

PBTs easily disperse in the atmosphere. When they fall or precipitate out of the atmosphere, they accumulate in water ecosystems and settle into sediment. Air, water, and sediment provide methods of exposure.

(continued)

- Since 1976, the Environmental Protection Agency (EPA) banned a total of six substances, chemicals, or chemical compounds.
- By March 2021, the EPA plans to ban an additional five PBTs [2].
- In 2001, the Stockholm Convention identified a dozen persistent organic pollutants (POPs) for ban by 2004.
- By 2019, the Stockholm Convention List includes 29 pollutants and related chemicals [3].
- The United States has not signed on to the Stockholm Convention.

Exposure Through Inhalation: Off-Gassing

Volatile organic compounds (VOCs) are chemicals that easily become vapors or gases to create toxic emissions. VOCs are inhaled and their emissions affect indoor air quality. They can also cause exposure through dermal contact. Exposure to VOCs for short periods of time can cause irritation, headache, dizziness, and affect vision. Long-term exposure can affect internal organs and cause cancer.

No federal regulations exist for non-industrial conditions, so the selection of low VOC materials is voluntary. The American Society of Heating, Refrigerating and Air-Conditioning Engineers (ASHRAE) provides direction on choosing materials with low emissions in its Indoor Air Quality Guide [4]. Low-emitting materials provide credits toward certification in Leadership in Energy and Environmental Design (LEED). In some cases, standards and certifications have adopted the California Department of Public Health Standard Method (V1.1, 2010) for measuring emissions. This is an example where voluntary standards adopt state regulation as a basis for testing or setting limits for exposure to earn certification (further explained in Chap. 10).

Exposure Through Inhalation: Particulate Matter (PM)

Particles suspended in air are defined as particulate matter. These particles can be solids or liquids and are small enough to inhale and lodge within the lungs. From the lungs they may move into the bloodstream. Their effects include aggravating asthma, altering heartbeat, decreasing lung function, irritating or swelling airways, and otherwise irritating the respiratory system.

The particles can be either organic or inorganic. This includes dust, pollen, soot, and smoke. Coarse particulates result from organic material or the breakdown of inorganic material. Fine particulates result from chemical reaction.

For reference, a human hair measures 50–70 microns. Coarse particulates (such as dust, pollen, or mold) measure 10 microns. Fine particles resulting from combustion or chemical reactions like smog measure 2.5 microns [5].

8.4 Levels of Toxic Impacts

Levels of toxic impacts are based on the character and severeness of exposure. The established categories are (Table 8.1):

- Acute, the result of a single harmful dose [6]
- Chronic, the buildup of smaller damaging doses over time
- Irritant, with a dose resulting in discomfort that subsides [7, 8]

Chronic toxicity is measured up to levels at which negative effects appear or the most amount of exposure that shows no negative effect. Acute toxicity is often measured down to a lethal dose or the smallest amount of exposure that results in death. For exposure to people:

- The National Institute for Occupational Safety and Health (NIOSH), a division of the Centers for Disease Control and Prevention (CDC), recommends exposure limits [9].
- The OSHA, a division of the Department of Labor, enforces exposure limits.

Table 8.1 Exposure types, definitions, and examples

Exposure Type	Definition	Cause
Acute	Harmful effects after a single or repeated exposure to a toxic material during a short time period Effects are noticeable, including sudden death Exposure often occurs due to an error or accident, though it can occur when risks are not properly contained	Industrial explosion Industrial leak or spill Unintentional chemical reaction Inappropriate ventilation
Chronic	Harmful effects after repeated or prolonged exposure Can be lethal, though more likely to cause changes in growth rates, reproduction, or behavior Effects typically appear after a long period of exposure Exposures can occur within the workplace, at home, or within the environment	Emissions, such as off-gassing of VOCs from a product Particulates in the air Continual skin contact Drinking or eating a substance that has been in contact with toxins
Irritant	Causes reversible inflammation to skin, eyes, and mucus membranes Eye irritants are tested through direct application to an animal's eye Animal testing is 4-hour-long applications at varied concentrations Chemical is an irritant provided any damage is naturally reversible in 21 days It is more difficult to quantify testing periods for exposure to human mucous membrane or upper respiratory inflammation due to difficulty scaling animal testing to human biology	Emissions, such as off-gassing of VOCs from a product Particulates in the air Continual skin contact Exposure to an allergen Exposure to a chemical reaction that produces a toxin

- Safety Data Sheets (SDS) publish exposure limits.

For exposure in the environment, the EPA suggests testing limits. The EPA divides exposure into near-field and far-field. Near-field exposure occurs through the use of consumer and in-home products. Exposure depends upon everyday activity and happens either at work or within the home. The EPA's Stochastic Human Exposure and Dose Simulation tool provides data for this exposure predicting both aggregate and cumulative exposure while considering different use patterns including consideration of demographics [10].

Far-field exposure occurs when there is industrial release into the broader environment. This release can be an accidental or planned release of waste. The EPA evaluates far-field exposure.

Chronic Lead Poisoning in Ancient Rome

It has been said that in ancient Rome lead (Pb) plumbing that supplied water throughout the infrastructure of the city caused widespread chronic exposure to lead toxins. Often it was thought that these toxins decreased intelligence and hurried the Roman Empire's fall. Recently there was a study of lead in sediment downstream from the city of Rome in comparison with sediment in ancient Roman pipes unearthed during archaeological digs. The study revealed lead in ancient Roman homes contained up to 100 times the levels found in the spring water supply for the city. While measurable, it was not enough to cause chronic toxicity across the populace [11].

8.5 Effect of Toxins

Toxins, radiation, pathogens, and pharmaceuticals can adversely affect human health through acute or chronic exposure. For the purposes of this book, the primary focus is upon toxins from building materials and products. Toxins can be categorized as (Table 8.2):

- Carcinogens
- Mutagens
- Teratogens
- Reproductive toxins
- Endocrine disruptors

At the broader scale of ecology, these same categories of toxins can affect animals. Aquatic toxicity affects water-dwelling organisms and their aquatic environment, including sediment. Terrestrial toxicity affects land-dwelling organisms through:

- Bioaccumulation, which is toxicity accumulated in one animal
- Biomagnification, which is toxicity that builds up across a food chain

Table 8.2 Substances that have toxic effects on human health [12–15]

Toxic Substance	Definition, Classification, or Categories	Examples
Carcinogen	Hazardous substance or exposure which poses the risk of causing cancer Carcinogens do not always cause cancer – the probability of getting cancer varies widely Categories: The International Agency for Research on Cancer (IARC; World Health Organization): *IARC Group 1*: Carcinogenic to humans *IARC Group 2A*: Probably carcinogenic to humans *IARC Group 2B*: Possibly carcinogenic to humans *IARC Group 3*: Not classifiable as to its carcinogenicity in humans The US National Toxicology Program (NTP), National Institutes of Health (NIH), CDC, and the Food and Drug Administration (FDA) *Known* to be human carcinogens *Reasonably anticipated* to be carcinogens The American Cancer Society compiles a list from both classifications, including substances or exposures that go by more than one name	Formaldehyde (urea-formaldehyde resin in fiberboard and particleboard) Asbestos (roofing felt) Benzene (paint remover, caulking, adhesives, plywood, fiberglass) Cadmium (electroplating, solar panels, paint) Chromium (alloy for stainless steel, coatings, preservative for structural timber) Nickel (alloy for stainless steel, nickel plating) Silica dust (brick, concrete, tile, drywall) Vinyl chloride (PVC – sewer pipe, window frames, phthalates – flooring, imitation leather)
Mutagen	A mutagen is a chemical or an exposure, such as radiation, that promotes errors in the replication of DNA. These errors can cause mutations in DNA. Mutagens *may* cause cancer, but not always *Category 1*, shown to have an effect on humans *Category 2*, likely to have an effect on humans, based on testing of animals *Category 3*, inconclusive evidence due to insufficient testing results All mutagens damage genetic information and are genotoxic Not all genotoxic substances are mutagenic: they may damage DNA replication, or cause incorrect timing in event activation (such as growth cycles) while not damaging DNA overall	Bromine (dyes, flame retardant) Arsenic (pressure- treated wood) Benzene, cadmium, chromium, and nickel (see description in carcinogen)

(continued)

Table 8.2 (continued)

Toxic Substance	Definition, Classification, or Categories	Examples
Reproductive Toxin	A reproductive toxin may negatively alter: Reproductive [sexual] organs Endocrine system (including the thyroid and adrenal glands) Exposure may affect: Infertility Difficulty in conception Child development, including after birth The Globally Harmonized System of Classification and Labelling of Chemicals (GHS) separates reproductive toxicity from mutagens and carcinogens, even though these hazards may also impact fertility Reproductive toxins also affect animals and plant species	Perfluorooctanoic acid (PFOA) and perfluorooctanesulfonic acid (PFOS) (Teflon and stain-resistant coatings for carpets and upholstery) Polybrominated diphenyl ethers (PBDEs – flame retardants) Formaldehyde and phthalates (see description in carcinogen)
Teratogen	Teratogens cause birth defects or developmental defects in children	Polychlorinated biphenyl (PCBs – used in HVAC and electrical equipment) Lead (alloy to brass, acoustic dampening panels, flashing, solder) Mercury (fluorescent lamps, gauges, batteries)
Endocrine Disruptor	Endocrine disruptors change the production of hormones and their interaction with receptors Four categories include both drug-based and environmentally based substances or chemicals: Estrogenic Anti-estrogenic Androgenic Anti-androgenic Estrogenic and androgenic compounds behave similarly to sex steroids (estrogen and testosterone), whereas anti-estrogenic and anti-androgenic substances bind to receptors, blocking hormones themselves from binding and thereby preventing the receptors' function	Bisphenol A (BPA – polycarbonate, epoxy resins) Dioxins (byproduct of bleaching), perfluoroalkyl and polyfluoroalkyl substances (PFAS – textile coatings) Triclosan (some antimicrobial cleaning products) Polychlorinated biphenyl, phthalates, and polybrominated diphenyl ethers (see above)

8.6 Measuring Toxicity

The science community defines benchmarks and thresholds for evaluating exposure levels. These benchmarks are then regulated by governmental agencies. Additional voluntary thresholds and levels are recommended by organizations that set standards.

Chemical concentration in a material is measured in parts per million. This may be parts per million in a solid or liquid or in the air. It may be measuring a concentration in raw materials, a finished product, or in waste. Parts per million is defined as 1 milligram per liter or 1 milligram per kilogram. For example, the concentration of carbon dioxide in the atmosphere passed 400 ppm in 2015. SDS regulated by OSHA, require manufacturers to disclose reproductive toxins, carcinogens, and Category 1 mutagens at a threshold of 1000 ppm and all other toxic hazards at a threshold of 10,000 ppm [16]. SDS also require companies to report exposure limits in parts per million.

Parts per million can also describe the threshold to which a product's chemical inventory is reported. Inventorying describes what is in a material but does not evaluate limits or thresholds for exposure. Chemical inventorying is necessary to earn a Health Product Declaration (HPD) (described in Chap. 11). HPD recognizes two thresholds, 1000 ppm where the chemical is publicly disclosed and 100 ppm where the chemical does not need to be disclosed [17].

Different agencies and organizations focus on different aspects of exposure depending on how they evaluate risk (Table 8.3). Exposure limits identify at what level a worker can be exposed to a toxin without excess harm. Often these limits

Table 8.3 Exposure limits [9, 19]

Exposure Type	PEL Permissible Exposure Limit	REL Recommended Exposure Limit	TLV Threshold Limit Value
Definition	Legal limit of exposure	Maximum level of exposure permitted	Limit of exposure with no adverse effect
Exposure Time	8 hours	10 hours	8 hours/day full length of one's career
Health/Cost	Health benefits and industry cost	Health benefit and industry cost	Health benefits alone
Organization or Agency	OSHA (Occupational Safety and Health Administration)	NIOSH (CDC) (National Institute for Occupational Safety and Health)	ACGIH (American Conference of Governmental Industrial Hygienists)
Impact	Legislated/regulated	Recommendation to OSHA for updates to PEL legislation	Recommendation
Updating Procedure	Many PELs not updated since the 1970s	Updated regularly	Updated yearly

date back to 1974. They originate from the 1970 Occupational Safety and Health Act and the 1977 Federal Mine Safety and Health Act [18].

Persistence and bioaccumulation evaluate how long a toxin will remain in an organism, a place, or an ecosystem. While exposure limits look to define a consistent quantity of exposure approved for workers, persistence and bioaccumulation focus upon the rate of decrease or increase in a concentration of a toxin (Table 8.4).

Dosage considers what concentration a toxin will cause irreversible harm, including death (Table 8.5). Testing methods for dosage were developed in 1927 and are used for solid, liquid, or aerosol/gas substances. Because it is unethical to test human subjects for lethal dosing, upper dosage values are arrived through:

- Animal testing
- Human cell study

Table 8.4 Persistence of effects of exposure

Persistence and Accumulation	Definition	Example
Half-life (T1/2)	A half-life is the time for a quantity to reduce to half of its original value by: Concentration Mass Radioactive decay Reactivity This rate can be affected through metabolism, accumulation in fatty tissues, and other hard to measure interactions	If an off-gassing chemical's half-life is a month in a building, strategies should ensure adequate fresh air for construction worker safety If a product's half-life is 10 years, then strategies should ensure how fresh air intake and ventilation through air conditioning and heating systems will provide adequate protection to office workers
Persistence	T1/2 of a chemical's quantity decreases in water, soil, sediment, or air	If 30 grams of a chemical with a half-life of 6 months is released into soil, it would result in the presence of 15 grams in 6 months, with the remaining half of the substance degrading into byproducts
Bioconcentration	Chemical accumulation in or on an organism Dosed through water, in relationship to the amount of the chemical in the water itself	Rain washes pesticides from crops into riverways and collects in fish that are later eaten by humans The resulting bioconcentration of chemicals can be measured in the fish or the humans
Bioaccumulation	Chemical accumulation in or on an organism either because uptake is faster than usability or chemicals cannot be broken down Uptake through land, air, or water	Emissions can accumulate in birds and other mammals near industrial manufacturing

Table 8.5 Dosage levels and effects on human health [20, 21]

Dosage	Effect	Substance	Definition
LD 50 Lethal Dose 50	Death Single dose Acute exposure	Solid, liquid, or aerosol/gas Toxin, radiation, pathogen	Median concentration of a substance that causes death in 50% of test animals during a specified duration of time
LD LO Lethal Dose Lowest Dosage	Death Acute or chronic exposure	Solid, liquid, or aerosol/gas Chemical of drug	Lowest dosage reported to kill animals or humans
LC 50 Lethal Concentration 50	Death Dose 4-hour exposure, 14 days' observation	Suspended in air: toxic gases, dusts, vapors, or mists Toxin	Median concentration that kills 50% of test animals
NOAEL No Observable Adverse Effect Level	No negative effect single dose	Drug, toxin, radiation	Highest dose of a substance or exposure without adverse effects which differs from the lowest dose for which there was an observed toxic or adverse effect

In both cases, tests assume scaling-up the dosage to a standardized person and do not account for preexisting conditions or variation in physiology. Lethal dosage typically focuses upon a single dose and does not take into consideration how a subject might react to an exposure that continues over a period of time. Lethal testing typically uses 50% as a benchmark, meaning that the dosage results in the death of half of the test subjects. A 50% benchmark reduces the testing of animals and leads to fewer animal deaths.

8.6.1 Evaluating Chemicals for Toxicity and Controlling Their Use and Release

With chemicals and their compounds proliferating since the middle of the twentieth century, a number of organizations developed systems to organize these substances. Today the International Union of Pure and Applied Chemistry Chemical Group and the CASRN system are most widely used by manufacturers and organizations [22].

Chemicals within these systems are banned, limited, or reduced through government regulation or through specific requirements of voluntary standards established by non-governmental organizations (NGOs) (Table 8.6).

Table 8.6 Chemical naming and numbering conventions [23]

Name	Originating Organization	Categories	Ordering Characteristics
Chemical Family Before 1990	The International Union of Pure and Applied Chemistry	Roman numeral (I–VIII) with A or B resulting in 18 families described: lithium, beryllium, scandium, titanium, vanadium, chromium, manganese, iron, cobalt, nickel, copper, zinc, boron, carbon, nitrogen, oxygen, fluorine, helium or neon Also categorized into four families: metals, metalloids, semimetals, non-metals	Similar chemical behaviors
Chemical Group Since 1990	The International Union of Pure and Applied Chemistry	18 numerical subgroups	Similarity of their outermost electron shells
CAS Numbering System	Chemical Abstracts Service	Roman numeral (I–VIII) with A or B Same categories as chemical family but numbers differ	Similar chemical behaviors
CASRN, Chemical Abstracts Service Registry Number	American Chemical Society	Unique numeric identifier 160 million chemical substances and 70 million protein and nucleic acid sequences	Numbers have no relationship to chemical structure or behavior Chemicals that have the same composition, but multiple names have a unique identifier

Individual chemicals and chemical families can be evaluated for toxicity based on:

- Regulation: Regulation provides subsets of substances that are banned or limited.
- Behavior: Certain chemical behavior results in consistent exposure risk to human health and ecology.
- Structure: The structure of certain chemicals and substances lead to risk based on chemical reaction.
- Use: The way in which certain chemicals and substances are employed increases their risk of harming humans and the environment.

Governmental agencies and non-governmental organizations establish chemical bans, restrictions, and reductions across several categories: specific chemicals, red lists, chemical categories, and chemical classes.

8.6.2 Bans on Specific Chemicals

The Toxic Substances Control Act (TSCA), the Toxics Release Inventory (TRI), and the Stockholm Convention on Persistent Organic Pollutants (POPs) define and categorize controlled and banned chemicals.

In 1976, Congress enacted the TSCA. At the time, over 62,000 chemicals were used in manufacturing. The act grandfathered tens of thousands of chemicals considered as not having "unreasonable risk," studied 200 chemicals, and banned only a handful of substances [24].

Chemicals banned in 1976 were:

- PCBs, used in electrical equipment, hydraulics, plasticizers, and fire retardants
- Chlorofluoroalkanes, used in aerosols
- Dioxins, a byproduct of bleaching paper pulp and burning waste
- Asbestos, an insulator, with some of the original ban overturned in 1991
- Hexavalent chromium, used in paint and coatings
- Nitrites, which mixed with certain salts or acids are used in manufacturing of metal parts

Additionally, the EPA regulates radon, lead, mercury, and formaldehyde. As of February 2020, the EPA also issued a preliminary decision to regulate production of perfluorooctanoic acid (PFOA) and perfluorooctanesulfonic acid (PFOS), such as stain-resistant coatings, nonstick coatings, and water repellents.

The TSCA places the burden of proof on the EPA to demonstrate that a chemical is toxic, rather than requiring that a manufacturer show that a chemical is not a hazard. Funding has never been sufficient for the EPA to cover the high cost of testing all materials. As a result, the agency relies on chemical companies to voluntarily provide testing data. In the 1990s, in an attempt to study the hazards of highly used chemicals, the EPA developed the High Production Volume (HPV) Challenge Program, asking chemical companies to voluntarily provide data on chemicals produced or imported measuring 1 million pounds or more [25]. Today over 82,000 chemicals are in production, and a searchable database on the EPA website names and describes nearly 4,300 HPV chemicals [26].

In 2016, the Frank R. Lautenberg Chemical Safety for the 21st Century Act was adopted, amending the TSCA.

- Within 3.5 years of adoption, the law required the EPA to have 20 ongoing risk evaluations.
- As of March 2019, 40 chemicals were prioritized for risk evaluation.
- By August 2019, 20 became high-priority substances.
- By March 2021, five additional PBTs will be banned.

The EPA compiles data from automated chemical screening on living cells or proteins, with information available from:

- The National Toxicology Program
- The National Institute of Environmental Health Sciences

- The National Center for Advancing Translational Sciences
- The Food and Drug Administration (FDA)

While existing chemicals are grandfathered by the original TSCA, manufacturers must provide data on new chemicals, and if the EPA cannot verify behavior or toxicity through similar chemicals, manufacturers are required to test the substance [27].

The Toxics Release Inventory (TRI) provides information on toxic chemical releases and pollution prevention activities reported by both industrial and federal facilities, tracking the management of toxic hazards to both human health and the environment [28]. The TRI program covers 767 individually listed chemicals and 33 chemical categories, though these numbers fluctuate constantly. There are 16 PBT chemicals and 5 PBT chemical compound categories that are subject to TRI reporting under Section 313 of the Emergency Planning and Community Right-to-Know Act (EPCRA). Release information includes emission into the air, water, through energy recovery, or treatment, or disposal into a landfill. This information is reported annually.

The Stockholm Convention on Persistent Organic Pollutants, a United Nations treaty signed in 2001, originally controlled production, use, trade, release, and disposal of 12 POPs. One hundred and eighty-five countries adopted and ratified the Stockholm Convention. Malaysia, Israel, and the United States are not signatories. By 2019, 29 pollutants and related chemicals made up the list.

8.6.3 Chemical Red Lists

NGOs and governmental agencies have developed proprietary red lists to either ban, limit, or reduce chemicals during the production of pharmaceuticals, consumer goods, and/or building materials (Tables 8.7 and 8.8). Competition between standards and certifying organizations means that there are several lists. Often subsets of these lists are derived from regulations set in place by the EPA, CDC, European Union Commissioner for the Environment, and California Proposition 65 [29, 30].

8.6.4 Chemical Class

A chemical class is grouped by similarity of:

- Carbon range number – number of carbon atoms in a chemical
- Chemical structure
- Function or use
- Interaction of substances
- Physical properties
- Radiological character

Table 8.7 Chemical red lists

Chemical Red List	Description	Number of Chemicals
SAICM The Strategic Approach to International Chemicals Management	International policy for producing and using chemicals safely worldwide by 2020, based upon the Johannesburg Plan of Implementation Adopted by the United Nations Environmental Programme Governing Council in February 2006	High concern: 205 chemicals/minerals Restricted: 74 substances
REACH The Registration, Evaluation, Authorisation and Restriction of Chemicals	European Union Commissioner for the Environment and the European Chemicals Agency Consumer product, apparel, and shoe industry	Restricted: 71 substances
DTSC Department of Toxic Substances Control in California	Protect residents and the environment from the "hazardous effects of toxic substances by restoring contaminated resources, enforcing hazardous waste laws, reducing hazardous waste generation, and encouraging the manufacture of chemically safer products"	3172 candidate chemicals
PROPOSITION 65 California Safe Drinking Water and Toxic Enforcement Act	Prohibits certain chemicals from being introduced to water supply; it also requires public notification – through signage – when a hazardous chemical is present in a location, including inside a building	1013 chemicals that produce hazards that cause cancer, birth defects, or reproductive harm

- Reactivity – physical or biological decomposition or the formation of new substances
- Testing behavior

Using these similarities, the EPA and other government agencies use data from tested chemical categories to fill data gaps for new untested chemicals. For regulations and standards, using a class structure to ban, limit, or reduce certain classes of chemicals provides an important opportunity [35]. If a specific chemical is banned, a processor or manufacturer may replace that chemical with one that is very similar, and equally toxic, but outside of the banned list [36]. By banning, limiting, or reducing a class of chemicals that all have similar properties, it is far more difficult to trade one toxic substance for another.

Using classes of chemicals to target a group of similar toxic chemicals is an alternative to the more common approach of using chemical red lists to target individual chemicals for elimination or limitation. Often, a ban on an individual chemical can result in replacement by an equally toxic chemical alternative with a similar function and slightly different formulation. For example, bisphenol A (BPA) has been eliminated from consumer products like water bottles and baby products over the last decade in response to growing concerns regarding human health impacts.

Table 8.8 Chemical red lists specific to the building materials and building products sectors [31–34]

Chemical Red List	Description	Number of Chemicals
ILFI International Living Future Institute	In order to achieve Living Building Challenge certification, no banned chemicals from the ILFI Red List can be present in the project	17 banned chemicals 503 chemicals on ILFI watch list
C2C Cradle to Cradle	Version 3 (2012) defines chemicals as technological and biological A technological chemical may be used in production or be present in a material but inert, and therefore does not pose a human health hazard A biological chemical may escape into the biosphere by off-gassing, entering water or foodways, or contaminating surfaces touched by the skin	62 biological chemicals 37 technical chemicals Reported when presence in formulation is beyond 1000 ppm
Perkins + Will Precautionary List	List generated by an internationally recognized architecture firm Some of the substances comprise chemical families rather than individual chemicals	56 current substances 22 retired substances 5 substances of emerging concern
LEED Pilot Credit and V4	LEED Pilot Credit 11 created a list for Chemical Avoidance in Building Materials in 2010 LEED Version 4 can no longer earn points using the Pilot Credit Instead LEED Version 4 favors Health Product Declaration, where disclosure of chemical makeup is required	15 substances or chemical families for Pilot Credit 11
The Pharos Database	Pharos is a database developed by the Healthy Building Network, a non-profit founded in 2000	160,000 substances evaluated for hazards, use, and exposure restrictions

These products are now advertised as "BPA-free." Unfortunately, one common replacement to BPA is bisphenol S (BPS), which has similar human health impacts to BPA. Raising awareness about classes of chemicals such as bisphenols rather than individual chemicals can avoid this type of regrettable substitution.

The "Six Classes Approach to Reducing Chemical Harm" was developed by the Green Science Policy Institute to help consumers, processors, manufacturers, and businesses avoid substituting toxic chemicals with chemicals of equally high human health impact [37] (See Chap. 12).

8.6.4.1 Per- and Polyfluoroalkyl Substances

Per- and Polyfluouralkyl substances (PFAS), also referred to as highly fluorinated chemicals or PFCs, are found in building materials that advertise resistance to water, oil, or stains, such as carpet, furniture textiles/coverings, and coatings or sealants. As noted in Table 8.2, PFAS are known reproductive toxins. They also have negative environmental impacts and are persistent in the environment, meaning they may travel and have impact over long distances without being diminished.

8.6.4.2 Antimicrobials

Antimicrobials inhibit growth and kill hazardous microbes. They are pesticides. Over the last decade, greater focus has been on how antimicrobials affect humans. In 2016, the FDA stopped the use of some antimicrobials in consumer products like hand soap and body wash to reduce bacterial resistance, but similar regulations have not been imposed on building materials [38].

Common antimicrobials include triclosan, triclocarban, nano-silver, and quaternary ammonium salts (quats). These antimicrobial chemical coatings and finishes are commonly applied to:

- Textiles and fabrics
- Door handles
- Sink handles
- Light switches
- Paint

Negative impacts on human and ecosystem health include the following:

- Too much skin contact leads to mutation of microbes to superbugs harmful to humans.
- Exposure causes the same hazards as to exposure to pesticides, metals, and hazardous chemicals including triclosan, nano-silver, and formaldehyde.
- Antimicrobials are toxic to aquatic organisms. They do not break down with wastewater treatment.

8.6.4.3 Flame Retardants

These chemicals are applied to materials to prevent the spread of flames and ignition. Flame retardants are applied to a range of furnishings and building materials, such as furniture foam, building insulation, textiles, paints, and wire sheathing. Alternatives to flame retardants are available for most use cases, such as smoking-free environments, sprinkler systems, and smoke detectors, all of which have been proven to be more effective at preventing flames.

Flame retardants are often sprayed on at the end of fabrication or applied to materials that gradually break down with wear, resulting in their travel into the air and in dust. As noted in Table 8.2, flame retardants are endocrine disruptors, teratogens, carcinogens, and reproductive toxins. They can also affect neurology and the immune system and are a particularly high risk for children [39] (See Chap. 12).

8.6.4.4 Bisphenols and Phthalates

Bisphenols and phthalates are widely used in plastics for a variety of performance characteristics, such as strength and flexibility. Bisphenols are most commonly found in polycarbonate plastic products like food containers, whereas phthalates are found in polyvinyl chloride (PVC) products like vinyl flooring and glues, caulks, paints, and air fresheners.

These chemicals have a range of health and environmental impacts, including asthma and neurodevelopmental problems in children and obesity, diabetes, heart disease, and reduced fertility in adults.

8.6.4.5 Some Solvents

Solvents include a large range of chemicals that are used to dissolve or disperse another chemical substance, including aromatic hydrocarbon solvents (toluene, xylene, benzene, and many others) and halogenated organic solvents (methylene chloride, perchloroethylene, trichloroethylene).

Solvents are found in a large range of wet-applied building products, such as oil-based paints, paint strippers, adhesives, finishes and sealants, and cleaning products. Chemical solvents are often carcinogenic. Many safer water-based alternatives exist.

8.6.4.6 Certain Metals

Mercury, arsenic, cadmium, lead, and naturally occurring metals are released into the soil, air, and water through fossil fuel combustion and industrial processes. Mercury, cadmium, and lead are also used directly in building products, such as fluorescent lighting (mercury), metal plating (cadmium), and piping (lead).

These metals harm brain development in children and increase the risk of cancer in adults. They have a range of negative effects on aquatic ecosystems and soil health, such as reduced reproduction rates and decreased growth of plants and animals.

8.6.5 *Characterizing, Optimizing, and Managing Chemicals*

Increasingly manufacturers are asked to disclose the chemical substances used in their processes and in their final products. Manufacturers may not retain a full ingredient list, especially if they purchase commodities or if their suppliers wish to retain their trade secrets. Chemical characterization allows a manufacturer to have their material tested after fabrication, providing a full list of all chemicals that are present in the final product. This characterization does not include chemicals that were used during steps in processing. In some cases, the chemical characterization can be used for sustainable certification, or it can be released publicly.

A benefit of knowing the ingredient list for a product is that a manufacturer can consider replacing toxic substances. Optimizing formulation allows the manufacturer to evaluate the chemical characterization of a product or material and then analyze that makeup based upon optimization for effectiveness or for concerns about toxicity. By going through this process, a manufacturer can find alternatives and options that can replace toxic substances but provide the same level of performance.

The collection of chemical formulation can be done by a manufacturer or by a third-party organization. When doing a toxicology assessment, contracting with an independent third-party organization means that objective data is collected at a facility that follows accepted testing practices. A third-party assessment is often required when seeking certification to a building materials sustainability standard.

Manufacturers are required to adhere to OSHA regulations as published through SDS with regard to the handling of individual toxic substances at individual points along the manufacturing process. A manufacturer can also develop and publish a holistic chemical management plan for hazardous chemicals on site. A chemical management plan can take into account all the steps of processing and manufacturing, to ensure that no systemic hazards exist. It can be very useful in risk management analysis. A manufacturer may also require public disclosure of similar plans from their suppliers. OSHA provides a Process Safety Management of Highly Hazardous Chemicals Standard, and also publishes permissible exposure limits, but does not require manufacturers to make publicly available a company's management plan for chemicals that are used on site at a specific facility.

References

1. ACS. (2020). *Hazard versus risk*. https://www.acs.org/content/acs/en/chemical-safety/basics/hazard-vs-risk.html. Accessed 14 Sept 2020.
2. United States Environmental Protection Agency. (2019). *EPA meets another Lautenberg deadline: Proposes persistent, bioaccumulative, and toxic chemicals rule under the toxic substances control act*. https://www.epa.gov/newsreleases/epa-meets-another-lautenberg-deadline-proposes-persistent-bioaccumulative-and-toxic. Accessed 14 Sept 2020.

3. United Nations Environment Programme. (2004). *Overview*. Stockholm Convention. http://www.pops.int/TheConvention/Overview/tabid/3351/Default.aspx. Accessed 14 Sept 2020.
4. ASHRAE. (2010). *Indoor air quality guide*. https://www.ashrae.org/technical-resources/bookstore/indoor-air-quality-guide. Accessed 14 Sept 2020.
5. United States Environmental Protection Agency. (2020). *Particulate matter (PM) basics*. https://www.epa.gov/pm-pollution/particulate-matter-pm-basics. Accessed 14 Sept 2020.
6. Interactive Learning Paradigms, Inc. (2020). *Acute toxicity*. MSDS HyperGlossary. http://www.ilpi.com/msds/ref/acutetoxicity.html. Accessed 14 Sept 2020.
7. Interactive Learning Paradigms, Inc. (2020). *Irritant*. MSDS HyperGlossary. http://www.ilpi.com/msds/ref/irritant.html. Accessed 14 Sept 2020.
8. MSDS Online. (2019). *Eye irritant*. MSDS Glossary of Terms. https://www.msdsonline.com/resources/sds-resources/glossary-of-terms/eye-irritation/. Accessed 14 Sept 2020.
9. United States Department of Health and Human Services, Centers for Disease Control and Protection. (2016). NIOSH Pocket Guide to Chemical Hazards. https://www.cdc.gov/niosh/npg/pgintrod.html. Accessed 14 Sept 2020.
10. United States Environmental Protection Agency. (2020). *Stochastic Human Exposure and Dose Simulation (SHEDS) to estimate human exposure to chemicals*. https://www.epa.gov/chemical-research/stochastic-human-exposure-and-dose-simulation-sheds-estimate-human-exposure. Accessed 14 Sept 2020.
11. Sumner, T. (2014). *ScienceShot: Did lead poisoning bring down ancient Rome?* Science, American Association for the Advancement of Science. https://www.sciencemag.org/news/2014/04/scienceshot-did-lead-poisoning-bring-down-ancient-rome. Accessed 14 Sept 2020.
12. American Cancer Society. (2019). *Known and probable human carcinogens*. https://www.cancer.org/cancer/cancer-causes/general-info/known-and-probable-human-carcinogens.html. Accessed 14 Sept 2020.
13. Department of Human Health and Services, National Institutes of Health, National Genome Human Research Institute (2020) *Mutagen*. https://www.genome.gov/genetics-glossary/Mutagen. Accessed 14 Sept 2020.
14. Interactive Learning Paradigms, Inc. (2016). *Reproductive toxin*. MSDS HyperGlossary. http://www.ilpi.com/msds/ref/reproductivetoxin.html. Accessed 14 Sept 2020.
15. Department of Human Health and Services, National Institutes of Health, National Institute of Environmental Health Sciences. (2020). *Endocrine disruptors*. https://www.niehs.nih.gov/health/topics/agents/endocrine/index.cfm. Accessed 14 Sept 2020.
16. Google. (2020). *Portico glossary*. https://support.google.com/healthymaterials/answer/6156109?hl=en. Accessed 14 Sept 2020.
17. HPD Collaborative. (2017). Using HPD 2.1 for USGBC's LEED v4 Material Ingredients Credit. https://www.hpd-collaborative.org/wp-content/uploads/2017/05/HPDC-v2.1-in-LEED-Basic-05-11-2017-final.pdf. Accessed 14 Sept 2020.
18. United States Department of Labor, Occupational Safety and Health Administration. (2016). *Hazard communication hazard classification guidance for manufacturers, importers, and employers*. OSHA 3844-02 2016. https://www.osha.gov/Publications/OSHA3844.pdf. Accessed 14 Sept 2020.
19. United States Department of Labor, Occupational Safety and Health Administration. (2016). *Chemical hazards and toxic substances*. https://www.osha.gov/SLTC/hazardoustoxicsubstances/. Accessed 14 Sept 2020.
20. Interactive Learning Paradigms, Inc. (2020). LD-50, 50% lethal dose. MSDS HyperGlossary. http://www.ilpi.com/msds/ref/ld50.html. Accessed 14 Sept 2020.
21. Dorato, M., Englehardt, J. (2005). *The no-observed-adverse-effect-level in drug safety evaluations: Use, issues, and definition(s)*. Abstract. National Library of Medicine, NIH. https://pubmed.ncbi.nlm.nih.gov/15979222/. Accessed 14 Sept 2020.

22. United States Department of Health and Human Services, Centers for Disease Control and Protection, Agency for Toxic Substances and Disease Registry. (2011). *Toxic substances portal*. https://www.atsdr.cdc.gov/substances/indexAZ.asp. Accessed 14 Sept 2020.
23. Interactive Learning Paradigms, Inc. (2020). CAS (chemical abstracts service) registry number. MSDS HyperGlossary. http://www.ilpi.com/msds/ref/cas.html. Accessed 14 Sept 2020.
24. United States Environmental Protection Agency. (2020). *EPA: High production volume list.*. https://comptox.epa.gov/dashboard/chemical_lists/EPAHPV. Accessed 14 Sept 2020.
25. Harrington, R. (2016). The EPA has only banned these 9 chemicals — Out of thousands. *Business Insider*. https://www.businessinsider.com/epa-only-restricts-9-chemicals-2016-2. Accessed 14 Sept 2020.
26. Perkins & Will (2020) Why material health? *Transparency*. https://transparency.perkinswill.com/about#governmentrole. Accessed 14 Sept 2020.
27. Lee Philips, M. (2006). Obstructing authority: Does the EPA have the power to ensure commercial chemicals are safe? *Environmental Health Perspectives, 114*(12) https://www.ncbi.nlm.nih.gov/pmc/articles/PMC1764141/.
28. United States Environmental Protection Agency. (2020). *Toxics release inventory (TRI) program*. https://www.epa.gov/toxics-release-inventory-tri-program. Accessed 14 Sept 2020.
29. Chem Safety Pro. (2019). REACH Restricted Substances Finder (RRS Finder). https://www.chemsafetypro.com/Topics/EU/REACH_Restricted_Substances_List_RRS_Finder.html. Accessed 14 Sept 2020.
30. State of California, Department of Toxic Substances Control. (2020). *Government links chemical information*. https://dtsc.ca.gov/scp/chemical-information/. Accessed 14 Sept 2020.
31. International Living Future Institute. (2020). *The red list*. https://living-future.org/declare/declare-about/red-list/#watch-list. Accessed 14 Sept 2020.
32. Cradle to Cradle. (2012). *Cradle to cradle certified banned list of chemicals*. https://www.c2ccertified.org/resources/detail/cradle-to-cradle-certified-banned-list-of-chemicals. Accessed 14 Sept 2020.
33. Perkins & Will. (2020). *Precautionary list*. https://transparency.perkinswill.com/lists/precautionary-list. Accessed 14 Sept 2020.
34. Healthy Building Network. (2020). *Pharos*. https://pharosproject.net/. Accessed 14 Sept 2020.
35. United States Department of Health and Human Services, Centers for Disease Control and Protection, Agency for Toxic Substances and Disease Registry. (2008). *Chemical classifications (chemicals according to their structure, properties, or use)*. https://www.atsdr.cdc.gov/substances/ToxChemicalClasses.asp. Accessed 14 Sept 2020.
36. Organisation for Economic Co-operation and Development. (2014). *Grouping of chemicals: Chemical categories and read-across*. https://www.oecd.org/env/ehs/risk-assessment/groupingofchemicalschemicalcategoriesandread-across.htm. Accessed 14 Sept 2020.
37. Green Science Policy Institute. (2020). Six classes. sixclasses.org. Accessed 14 Sept 2020.
38. Perkins & Will. (2017). *Top 10 things you need to know about antimicrobials*. http://assets.ctfassets.net/t0qcl9kymnlu/5qx5zOxBbqIuSMmQEAIKMS/83ba8a8586a14411089142a839c02b70/Antimicrobial_WhitePaper_10Takeaways.pdf. Accessed 14 Sept 2020.
39. Department of Human Health and Services, National Institutes of Health, National Institute of Environmental Health Sciences. (2020). *Flame retardants*. https://www.niehs.nih.gov/health/topics/agents/flame_retardants/index.cfm

Chapter 9
Social Accountability

9.1 Introduction

Every environmental impact discussed in Chapters 4 through 8 affects individuals and communities. From extracting raw materials to climate change to pollution to toxicity, each impacts people. This chapter discusses the fair treatment of workers and the preservation of their communities as sustainable initiatives expressed through labor and human rights (Fig. 9.1).

Social accountability evaluates fair treatment through adherence to treaties, laws, and regulations that mandate equality. Additionally, social accountability encourages ethical business practice.

Social justice differs from social accountability. It goes beyond simple adherence to civil and criminal law, principles of economic supply and demand, and moral frameworks [1]. It is defined as seeking a society where allocations of both things of value and burden are distributed equitably – whether in individual interaction, through the acts of corporations and government, or at the broadest level of social structure. The UN's Human Development Index seeks to implement social justice [2]. Social justice does not just focus upon whether a person has the right to something, but whether their situation gives them the freedom of choice and capability to attain that thing of value or shoulder a burden [3]. This logic defines the fundamental difference between equality and equity. Social fairness is sometimes used as an interchangeable term. Laws and standards have not yet developed adequate models to evaluate if workers and communities are treated with equity. This chapter discusses frameworks that are currently in place, and the authors acknowledge that these frameworks do not properly address evaluation methods to achieve social justice.

Social accountability, as defined by labor and human rights, is established through international treaties and individual countries' laws such as the US civil rights laws. Standards recognized by the International Organization for Standardization (ISO) establish voluntary management structures and codes of conduct. Currently, social

H. R. Roth et al., *The Green Building Materials Manual*,
https://doi.org/10.1007/978-3-030-64888-6_9

INITIATIVES AND IMPACTS: SOCIAL ACCOUNTABILITY

Social Accountability

Labor and Human Rights:
Child and Forced Labor
Discrimination
Working Hours
Unions and Collective Bargaining
Management Processes

Social Impact Indicators
Safe Working Environments
Requirements for Grievance Mechanisms
Third Party Audit
Animal Welfare

Fig. 9.1 Social accountability initiatives and impacts

SOCIAL ACCOUNTABILITY DURING LIFE CYCLE

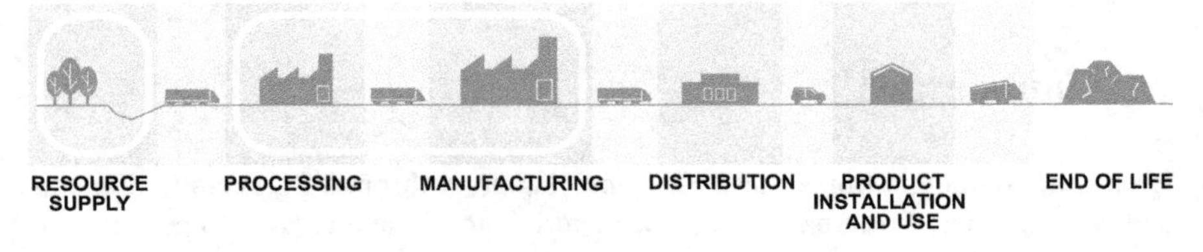

Fig. 9.2 Most emphasis on social accountability is on the resource supply, processing, and manufacturing stages of the life cycle

accountability impacts and initiatives focus upon the life cycle stages of resource material supply and extraction, material processing, and manufacturing (Fig. 9.2). Within the United States, labor and human rights during the installation stage of life cycle fall under labor laws and regulations. Beyond laws and ISO standards, SA8000, the International Living Future Initiative Social Equity Petal, and the JUST Label are voluntary standards that address social accountability.

Social Fairness

Material sustainability standards may seek statements, or commitments, to support or promote a culture of social fairness. Goals often encompass being diverse and inclusive in hiring strategies, fair in pay, understanding in performance evaluation, and offering flexible incentive structures.

9.2 Labor and Human Rights

In 1948, the United Nations proclaimed the Universal Declaration of Human Rights (UDHR) which established international human rights law. It was initially signed by 48 nations, with 8 abstentions. Since its proclamation, the UDHR has generated human rights treaties, conventions, provisions in constitutions, and laws within individual countries. The declaration celebrated its 70th anniversary in 2019 [4].

Articles 1 and 2 of the UDHR assert:

All human beings are born free and equal in dignity and rights...without distinction of any kind, such as race, colour, sex, language, religion, political or other opinion, national or social origin, property, birth or other status. [5]

Later in the document, Article 23 focuses upon a human being's right to work [5]:

1. Everyone has the right to work, to free choice of employment, to just and favorable conditions of work, and to protection against unemployment.
2. Everyone, without any discrimination, has the right to equal pay for equal work.
3. Everyone who works has the right to just and favorable remuneration ensuring for himself and his family an existence worthy of human dignity, and supplemented, if necessary, by other means of social protection.
4. Everyone has the right to form and to join trade unions for the protection of his interests.

There are currently nine major international human rights treaties within the UDHR. The United Nations Human Rights Office of the High Commissioners provides an index of which member states have signed on to which treaties. The United States has not signed on to treaties prohibiting torture, the death penalty, and protecting migrant workers' rights [6]. To enforce treaties at a governmental scale, the UN Security Council may:

- Send in military observers or peacekeeping forces
- Issue economic sanctions
- Issue embargos on weapons
- Level financial penalties
- Issue travel bans
- Sever diplomatic relations
- Institute blockades
- Institute military action against nations

At a corporate scale, the UN seeks a top-down approach asking business leaders to adopt the principles of the declaration. From a bottom-up approach, organizations such as Amnesty International publicize when corporations exploit badly enforced laws and violate human rights [7]. To achieve its mission, Amnesty International:

- Collects data
- Lobbies and organizes campaigns and protests
- Generates litigation based on the Universal Declaration of Human Rights

9.2.1 Assessing and Addressing Child Labor and Forced Labor

Child labor and forced labor are prohibited by the Universal Declaration of Human Rights (1948) and the UN International Covenant on Economic, Cultural and Social Rights (1966). Since 2009, the Bureau of International Labor Affairs within the US

Department of Labor has issued a "List of Goods Produced by Child Labor or Forced Labor" [8]. While not listing specific companies or manufacturers, this report specifies goods and countries where child labor practices exist. Cotton, making bricks, and mining practices lead the list. In 2018, the list included 148 goods from 76 countries.

Child labor is defined as interfering with schooling, and "affecting health and development including activities that are mentally, physically, socially, or morally dangerous [9]".

Forced labor occurs when "individuals are compelled to provide work or service through the use of force, fraud, or coercion" [10]. Forced labor is also referred to as unfree labor. Forced labor can include [11]:

- Withholding workers' documents to restrict movement and ability to change jobs
- Forcing excessive overtime
- Abuse in working and living conditions
- Abuse of a worker's vulnerabilities
- Isolation of the worker
- Violence to the worker
- Withholding pay until some requirement is met
- Prison labor

Voluntary standards ask that a company assess work completed during stages of resource extraction and manufacturing to verify that no child or forced labor exists. This is of particular concern in processes completed in developing countries, or where information on supply chain cannot be traced. The International Labour Organization (ILO; a division of the UN) and the Organisation for Economic Co-operation and Development (OECD) work to set labor policy. The International Organization for Standardization (ISO) sets standards for employment structures and management of workers.

9.2.2 Discrimination

In terms of labor and the workplace, discrimination is allowing gender, race, age, disability, or religion to affect how two equally qualified individuals are treated in the workplace. Title VII of the 1964 Civil Rights Act prohibits discrimination in companies with more than 15 employees where employment lasts more than 20 weeks a year [12]. Some standards and certifications provide language requiring that facilities or suppliers outside the United States adhere to Title VII legislation. Title VII provides grievance mechanisms, described later in this chapter, to allow workers to file complaint against their employer for conditions of discrimination. Evaluation of grievances can lead positively to changes in workplace treatment or can become the basis for litigation.

9.2.3 *Working Hours*

The International Labour Organization (ILO) along with the United States and many other countries, mandate a 40-hour work week for hourly employees. However, some employers worldwide require excessive work hours per day or across a work week. Additionally, employers may require workers to labor for periods of time without pay.

Employers in the United States must adhere to the Fair Labor Standards Act [13]. However, within the United States, some employers circumvent the act by misclassifying workers as independent contractors or classifying non-exempt employees as exempt. This means that workers may be required to work beyond an ILO-mandated work week, or may not be provided overtime pay for hours worked [14].

9.3 Unions and Collective Bargaining

Across the globe, employees are often not allowed to negotiate as a group with employers. This puts fair and equitable practices at risk [15].

In 1935, the US National Labor Relations Act established workers' rights to form independent labor union organizations, to collectively bargain, and to strike. The act also prevents unions from engaging in practices that are unfair either to employers or employees. When the National Labor Relations Act was voted into law, it overturned a patchwork of courtroom judgments that prohibited independent organizations from bargaining for employees or encouraging strikes. It also established the National Labor Relations Board to bring violations of labor law to court. Passage of this act occurred during the height of the Great Depression and signified governmental steps to exert control over private industry by allowing representatives to intermediate between citizens and business.

Unions maintain workers' rights by protecting wages, hours, and working conditions. They also often provide pensions and healthcare. Instead of legislating workers' rights, union-based labor negotiates workers' rights. In other countries, unions provide additional protections such as unemployment insurance [16].

9.4 Management Processes and Social Accountability

A range of management processes exists to value the health and welfare of workers during business operations. ISO has developed families of voluntary standards that provide both guidance and organizational structures for monitoring manufacturing processes (Table 9.1). Many other standards reference this ISO framework as the established way to organize the implementation of social responsibility. ISO frameworks are adopted in a top-down approach, with leadership providing the management for the implementation of these structures across their facilities and employees.

Table 9.1 ISO standards with their purposes defined

Standard (Or Family of Standards)	Outcome	Description
ISO 26000 [17]	Guidance for companies to make statements and develop goals regarding social responsibility	Principles: accountability, transparency, ethical behavior, respect for stakeholder interests, respect for rule of law, respect for international norms of behavior, respect for human rights Subjects: organizational governance, human rights, labor practices, environment, fair operating practices, consumer issues, community involvement, and development The standard does not provide any metrics or methods by which to enact the statements
ISO 19600 [18]	Guidelines to establish management policies to comply with risk management	Establish guidelines to evaluate risk to workers, operations, and the environment, and outline the establishment of methods to comply with rules to limit risk
ISO 14000 [19]	Establish management systems to minimize environmental impact	Minimize environmental impact while adhering to laws and regulations
ISO 9000 [20]	Provide quality management systems	Establish principles to ensure adherence to laws and regulations for quality control, accounting disclosure, and employee training Manufacturers often require that their suppliers also use these standards to ensure they adhere to laws and regulations

9.4.1 Social Impact Indicators

Some certifications and standards seek metrics on a range of social impact indicators. The CDC defines indicators as the measurable data that allows an organization to know if their work is achieving planned outcomes [21].

- Input indicators measure what is necessary to implement a project.
- Process indicators verify if a project is implemented as planned.
- Impact or outcome indicators measure whether outcomes are having expected effects.

Social impact indicators communicate clear measurable outcomes for a field with many goal-oriented statements. For example, if a goal is "to seek social justice," indicators should be communicated in order to measure success in direct correlation with the goal in order to produce data that can be evaluated.

9.4.2 Risk Assessment

Some standards and certifications ask companies to provide risk assessment at their own facilities or at those of their suppliers. Assessing risk quantifies the likelihood that behavior will result in harm. Corporations weigh both ethical responsibility, and the cost of litigation when making decisions on behavior. The EPA defines two types of risk assessment: human health risk assessment and ecological health risk assessment [22].

- Human health assesses risk associated with the probability and type of exposure of chemicals or other contaminants that may occur to employees or the broader community.
- Ecological risk assesses the environmental impact of exposure to chemicals, disease, climate change, or invasive species that might result from manufacturing processes.

9.4.3 Safe Working Environments

Establishing a safe workplace focuses upon lessening the risk of dangers to worker health. These dangers can be physical dangers or perceived threats. Evaluating a working environment for safety can include:

- Equipment safety
- Chemical handling
- Indoor air quality
- Housing (if provided)
- Social interaction
- Other physical safety issues related to a product's life cycle

Establishing a safe working environment includes adhering to regulations on how chemicals should be labeled and handled, developing reporting requirements for incidents and violations, and monitoring or auditing whether policies are enforced. For example, a manufacturer may be required to annually report the number of injuries in a facility. Data on these injuries can lead to an investigation which may change operations, decreasing risk to workers. Grievance mechanisms, described below, allow workers to file complaints about unsafe conditions. Evaluation of grievances can lead positively to changes in workplace protocols or can become the basis for litigation.

9.4.4 Requirements for Grievance Mechanisms

A grievance mechanism is an established structure where workers can [23]:

- File a complaint with confidentially
- Have their concern evaluated without prejudice

- Resolve concerns without threat of retribution

Typically, grievance mechanisms exist within a company and accommodate employee complaints. In addition to seeking to correct an individual grievance, steps should also be taken to ensure broader human rights infringements do not occur. Support for employees comes through:

- Trade unions
- Whistle-blower protection
- Reporting systems for sexual harassment
- Health and safety incident forms
- Complaints to management

Grievance mechanisms are required by civil rights law in the US. By contrast, communities often have less defined access to grievance mechanisms.

9.4.5 Third-Party Audit or Accreditation

Third-party audits for social accountability are similar in structure to independent audits for the other sustainability impact categories in this chapter. A company might develop a quality management system adhering to standards set by ISO.

An independent firm is then hired to assess the success of the system. Because it is an outside firm, this assessment is objective. If a company evaluates its own suppliers, it is providing a second-party audit; if a company internally assesses its social accountability, it is considered a first-party audit [24]. Both second- and first-party audits are considered more likely to be subjective as the company has a direct role in evaluation.

9.5 Animal Welfare

Some building materials, such as upholstery, and certain adhesives, may use animal products. Some standards may contain language protecting animal treatment. Principles regarding this treatment derive from language developed in 1965 by the government of the United Kingdom in response to animal husbandry. This language was formalized as the Five Freedoms in 1979 by the UK Farm Animal Welfare Council and was adopted by a variety of animal organizations and corporations. The Five Freedoms include [25]:

1. Freedom from Hunger and Thirst. By ready access to freshwater and diet to maintain health and vigor
2. Freedom from Discomfort. By providing an appropriate environment including shelter and a comfortable resting area

3. Freedom from Pain, Injury, or Disease. By prevention or rapid diagnosis and treatment
4. Freedom to Express Normal Behavior. By providing sufficient space, proper facilities, and company of the animal's own kind
5. Freedom from Fear and Distress. By ensuring conditions and treatment which avoid mental suffering

References

1. Chappelow, J. (2019). Social justice definition. *Investopedia*. https://www.investopedia.com/terms/s/social-justice.asp. Accessed 12 Sept 2020.
2. United Nations Development Program, Human Development Reports. (2020). Human Development Index (HDI). http://hdr.undp.org/en/content/human-development-index-hdi. Accessed 12 Sept 2020.
3. Robeyns, I. (2016). *The capability approach*. The Stanford Encyclopedia of Philosophy, Edward N. Zalta (ed.). https://plato.stanford.edu/entries/capability-approach/. Accessed 12 Sept 2020.
4. National Public Radio. (2018). Boundlessly Idealistic, Universal Declaration of Human Rights Is Still Resisted. https://www.npr.org/2018/12/10/675210421/its-human-rights-day-however-its-not-universally-accepted. Accessed 12 Sept 2020.
5. United Nations. (1948). *Universal declaration of human rights*. https://www.un.org/en/universal-declaration-human-rights/. Accessed 12 Sept 2020.
6. United Nations Human Right Office of the High Commissioner. (2020). Status of Ratification Interactive Dashboard. https://indicators.ohchr.org/. Accessed 12 Sept 2020.
7. Amnesty International. (2020). *Corporations*. https://www.amnesty.org/en/what-we-do/corporate-accountability/. Accessed 12 Sept 2020.
8. United States Department of Labor. (2018). US. Department of Labor's 2018 List of Goods Produced by Child Labour or Forced Labour. https://www.dol.gov/sites/dolgov/files/ILAB/ListofGoods.pdf. Accessed 12 Sept 2020.
9. International Labour Organization, International Programme on the Elimination of Child Labour. (2020). What is child labour? https://www.ilo.org/ipec/facts/lang%2D%2Den/index.htm. Accessed 12 Sept 2020.
10. United States Department of Homeland Security. (2020). What is Forced Labor? https://www.dhs.gov/blue-campaign/forced-labor. Accessed 12 Sept 2020.
11. International Labour Office. (2012). *ILO indicators of forced labour*. http://www.ilo.org/wcmsp5/groups/public/%2D%2D-ed_norm/%2D%2D-declaration/documents/publication/wcms_203832.pdf. Accessed 12 Sept 2020.
12. United States Equal Opportunity Commission. (1964). Title VII of the Civil Rights Act of 1964. (Pub. L. 88-352) (Title VII) Volume 42 of the United States Code, beginning at section 2000e. https://www.eeoc.gov/statutes/title-vii-civil-rights-act-1964. Accessed 12 Sept 2020.
13. United States Department of Labor, Wage and Hour Division. (2020). Handy Reference Guide to the Fair Labor Standards Act. https://www.dol.gov/agencies/whd/compliance-assistance/handy-reference-guide-flsa#1. Accessed 12 Sept 2020.
14. International Labour Standard. (2020). International Labour Standards on Working Time. https://www.ilo.org/global/standards/subjects-covered-by-international-labour-standards/working-time/lang%2D%2Den/index.htm. Accessed 12 Sept 2020.
15. International Labour Standard. (2020). International Labour Standards on Collective Bargaining. https://www.ilo.org/global/standards/subjects-covered-by-international-labour-standards/collective-bargaining/lang%2D%2Den/index.htm. Accessed 12 Sept 2020.

16. Matthews, D. (2017). Europe could have the secret to saving America's unions. Vox, Voxmedia. https://www.vox.com/policy-and-politics/2017/4/17/15290674/union-labor-movement-europe-bargaining-fight-15-ghent. Accessed 12 Sept 2020.
17. ISO. (2010). ISO 26000 social responsibility. https://www.iso.org/iso-26000-social-responsibility.html. Accessed 18 Jan 2021.
18. ISO. (2015-18). ISO 19600 compliance management systems - guidelines. https://www.iso.org/standard/62342.html. Accessed 18 Jan 2021.
19. ISO. (2015-19). ISO 14000 family - environmental management. https://www.iso.org/iso-14001-environmentalmanagement.html. Accessed 18 Jan 2021.
20. ISO. (2015-18). ISO 9000 family - quality management. https://www.iso.org/iso-9001-quality-management.html. Accessed 18 Jan 2021.
21. Centers for Disease Control and Prevention, Program Performance and Evaluation Office (PPEO). (2016). Indicators CDC Approach to Evaluation. https://www.cdc.gov/eval/indicators/index.htm. Accessed 12 Sept 2020.
22. United States Environmental Protection Agency. (2020). Risk Assessment. https://www.epa.gov/risk. Accessed 12 Sept 2020.
23. Vermijs, D et al. (2014). Chapter 3.8 Remediation and grievance mechanisms 'Early warning, effective solutions.' Doing Business with Respect for Human Rights Guide, Global Compact Network Netherlands, Oxfam, Shift. https://www.businessrespecthumanrights.org/en/page/349/remediation-and-grievance-mechanisms. Accessed 12 Sept 2020.
24. Hammar, M. (2015). *First-, Second- & Third-Party Audits, what are the differences?* ISO 9001 Blog. https://advisera.com/9001academy/blog/2015/02/24/first-second-third-party-audits-differences/. Accessed 12 Sept 2020.
25. Farm Animal Welfare Council (FAWC) (1979) *Five freedoms.*. https://webarchive.nationalarchives.gov.uk/20121010012427/http://www.fawc.org.uk/freedoms.htm. Accessed 12 Sept 2020.

Chapter 10
Laws, Regulations, Standards, Certifications, and Ecolabels

10.1 Introduction

Chapters 4, 5, 6, 7, 8, and 9 describe environmental initiatives and impacts that relate to building materials and products. This chapter explains how these initiatives and impacts are evaluated. It describes both the structures that exist for evaluation (Fig. 10.1) and who verifies that the evaluation is fair and true. The chapter answers two main questions:

- What legal structures establish and enforce required environmental protection?
- How can manufacturers show the ways in which they voluntarily protect the environment?

10.2 Laws and Regulations

The government uses laws to establish required rules of behavior and corresponding punishment if those rules are broken. Before a law is voted upon, it is called a bill. Afterward, it is called a statute or an act. Laws can be amended through a vote by the legislature. In some cases, laws are written to expire at a future date. Breaking environmental laws results in punishment including fines, cease and desist orders, and rarely incarceration. Laws relating to environmental and social impacts include (Fig. 10.2):

- International laws focusing upon broad global issues
- Federal laws that center upon issues that go across state boundaries
- State or municipal laws responding to regional environmental conditions or local issues of value

H. R. Roth et al., *The Green Building Materials Manual*,
https://doi.org/10.1007/978-3-030-64888-6_10

Fig. 10.1 Laws and standards

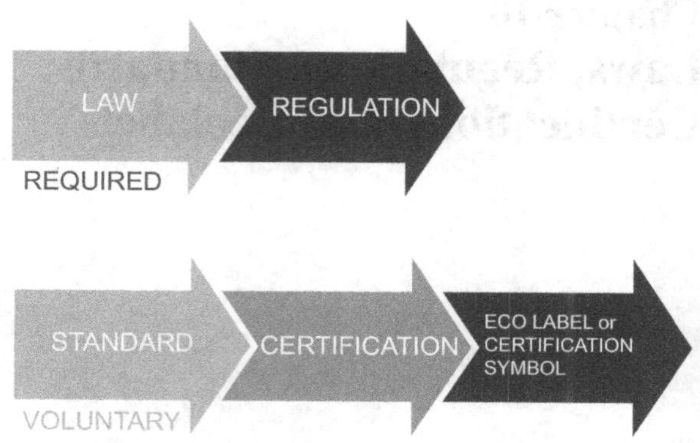

REGION	TYPE	DESCRIPTION	EXAMPLE
	International Law and Treaties	Focus upon broad global issues	Universal Declaration of Human Rights
	Federal Laws and Regulations	Issues that expand across state boundaries	The Clean Water Act
	Regional or State Laws, Regulations, and Standards	Regional environmental conditions or issues of value to a local area	California Specification 01350 for Indoor Air Quality

Fig. 10.2 Global, federal, and local jurisdictions

Regulations interpret a law. Regulations establish rules for monitoring and enforcing the law and how those rules are carried out. Regulations also establish the punishment if rules are not followed. In the United States, federal, state, or municipal agencies establish and enforce regulations.

For example, Congress enacted the Clean Water Act (CWA) in 1972. One part of the law states "It is the national policy that the discharge of toxic pollutants in toxic amounts be prohibited" [1]. The Environmental Protection Agency (EPA) is responsible for regulating the law by establishing the scope and thresholds for key questions created by the law:

- What are toxic pollutants?
- At what concentration are they toxic?

- What are the approved ways of testing?
- How is the data collected, formatted, and reported?

In addition to laws and regulations, governmental offices have other ways to monitor and reduce environmental impacts (Fig. 10.3).

10.3 Executive Orders

The President of the United States can issue an executive order. An executive order interprets the constitution or a law. It directs how a reading of the constitution or law might be expanded or reduced in scope. An executive order can also provide

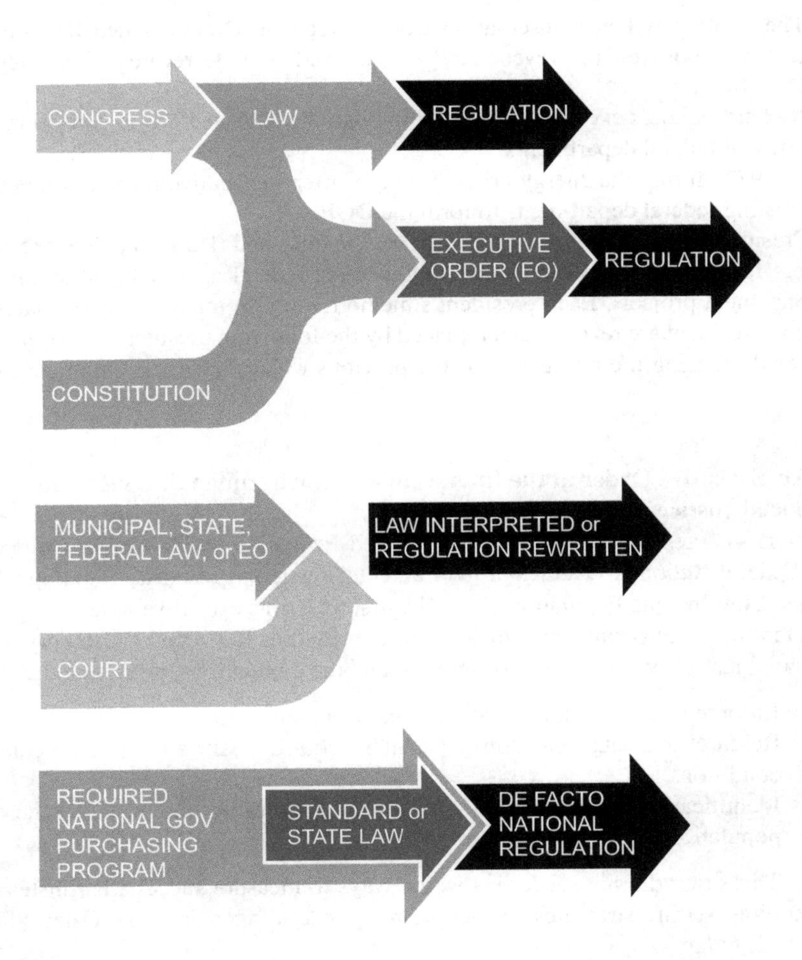

Fig. 10.3 Establishing regulations

instruction on enforcement and regulation. An executive order can be repealed or altered by future presidents, or it can be struck down by the court system. Federal agencies are created by executive order. Examples include the EPA and the Department of Energy (DOE). Federal agencies reside within the executive branch.

Between 1903 and 1908, Theodore Roosevelt instituted some of the most far-reaching environmental protections still in existence within the United States today. He issued executive orders that established:

- The National Wildlife Refuge System
- The National Forests
- National Monuments, as an interpretation of the Antiquities Act, which became the foundation of the National Park System

Additional executive orders that provided environmental protection include [2, 3]:

- The 1930s Civilian Conservation Corps Executive Orders issued by Franklin Delano Roosevelt to prevent land erosion and provide reforestation after the Dust Bowl
- Richard Nixon's executive order to establish the EPA in 1970 to consolidate an array of federal departments
- In 1977, during the energy crisis, Jimmy Carter's executive order to reorganize existing federal departments to form the DOE

Presidents Bush Sr., Clinton, Bush Jr., Obama, and Trump passed executive orders focusing on environmental initiatives for federal agency construction and management projects. Each president's methods vary. In many cases mandates set by one president are revoked and replaced by the following president either increasing or decreasing the stringency of the previous executive order. Often executive

An Executive Order at the Intersection of Environmental Issues and Social Justice

In 1994, President William Clinton issued Executive Order 12898, entitled "Federal Actions to Address Environmental Justice in Minority Populations and Low-Income Populations." For the first time, this executive order brought attention to environmental justice, an area within the broader category of social justice. Within minority and low-income populations, it promoted:

- Enforcement of existing health and environmental law
- Research and data collection on the intersection of health and environmental conditions
- Identification of different patterns of use of natural resources within these populations

This order did not set forth specific ways to measure success, but instead required setting strategies for adoption by federal agencies. The order still stands today [4].

orders set goals beyond the conclusion of the presidency, calling into question the effectiveness of this method of governing.

Executive Orders on Sustainability Across Federal Works and Their Consequences

The federal government is one of the largest consumers of goods and services in the country. Several presidents have used this buying power to increase sustainable initiatives and reduce environmental impacts across the building material and building product sectors.

In 1999, President Clinton enacted "Greening the Government through Efficient Energy Management," EO 13123. This order required that federal agencies reduce greenhouse gases, energy consumption, and water use by specific percentages. Additionally, this order required the General Services Administration (GSA) and the Department of Defense (DOD) enhance their preferred purchasing programs to ensure products and services met these new efficiency requirements. The executive order's efficiency requirements were to be completed by 2010. Requiring these savings by a date beyond the end of Clinton's presidency by nature put into question the attainability of the requirements, since often executive orders are revoked or replaced by the next presidency.

President Clinton's executive order was indeed revoked by President George Bush Jr.'s EO 13423, which replaced the deep percentage increases in efficiency with reduced benchmarks. Again, deadlines for completion were set beyond the conclusion of Bush's presidency.

In 2009, President Barack Obama issued EO 13514 Federal Leadership in Environment, Energy, and Economic Performance [5]. He then updated and replaced that executive order with EO 13693, Planning for Federal Sustainability in the Next Decade [6], in 2015 (revoked by President Trump's EO 13834 in 2018). President Obama's executive orders required that the federal government purchase "products and services that are energy-efficient, water-efficient, biobased, environmentally preferable, non-ozone depleting, contain recycled-content, non-toxic or less-toxic alternatives." These orders mandated that a percentage of federal projects meet sustainable guiding principles and that by 2020 all federal buildings entering the planning process meet 2030 zero-net-energy standards. Obama's EO 13693, which replaced EO 13514, further increased efficiency across energy and water while expanding renewable energy sources.

By contrast, President Trump's EO 13834 [7], in 2018, removed future goals. Instead, it required that purchasing and planning simply rely upon current statutes and regulations. This exhibits a fundamental difference in strategy: Obama's executive orders planned to move the market through regulation, whereas Trump's executive order follows the market. This difference clearly shows how executive order can expand or reduce control over private industry.

10.4 Court Challenges to the Constitution, Laws, Executive Orders, and Regulations and Case Law

The three branches of the government are the legislative, the executive, and the judicial. Court cases can lead to interpretation or revocation of laws and regulations set by Congress and the president. They test the constitutionality of law and the interpretation of law through regulation. They also challenge executive orders and regulations created by federal or state agencies. Additionally, court cases can lead to case law which sets precedent for rules and behavior. Case law is not voted upon by a government body but is accepted as binding and must be followed.

A case may begin with a private corporation challenging a state regulation. As it moves up through the courts, the case may end in the US Supreme Court challenging not only the constitutionality of the state regulation but the state law itself, while also creating implications for broader federal law. In environmental law, court cases can affect law and regulations focusing upon:

- Resource extraction
- Energy use
- Air pollution
- Water use, discharge, and pollution

The Case for Government Restricting Emissions

In 2007, the State of Massachusetts sued the EPA for failing to regulate car and truck tailpipe emissions, arguing that it was causing harm to the state. The Bush administration asserted that the Clean Air Act did not establish the right to regulate carbon emissions, so the EPA was not required to enact emissions regulations. The Supreme Court ruled that the EPA could not refuse to regulate carbon emissions unless it produced scientific evidence against climate change.

> Given EPA's failure to dispute the existence of a causal connection between man-made greenhouse gas emissions and global warming, its refusal to regulate such emissions, at a minimum, "contributes" to Massachusetts' injuries [8].

The Supreme Court's decision required the EPA regulate emissions limits on cars and trucks in the future. Additionally, this ruling signaled that the government had a broader responsibility in the future to address climate change through regulating emissions on factories and power plants [9, 10].

10.5 *De Facto* Regulations

De facto regulations set behavior not through law, but through motivation. Across the country, there are many large purchasing programs run by agencies ranging from the GSA to the DOD, to the California School Boards Administration.

Executive orders from the president or state governor currently require these agencies to use preferred purchasing programs when buying building materials and products.

These purchasing programs can influence private companies to increase their sustainability initiatives by requiring that products be environmentally certified or provide a certain level of energy or water efficiency. For example, these purchasing programs require that agencies specify energy-efficient and water-conserving equipment, in particular bearing the EPA "Energy Star" or "WaterSense" ecolabel. To be considered for GSA or DOD projects, products must *de facto* meet the requirements of "Energy Star" and "WaterSense" standards set by the EPA.

As another example, the California Department of Public Health sets standard practice for testing and benchmarking low-emitting materials for installation in California schools through Specification 01350. The California school system is a large buyer of interior materials, furniture, and fixtures. In order to be considered for purchase by California schools, materials and products must meet the requirements of Specification 01350 for low emission of volatile organic compounds. Because these same vendors sell products across the country, these products also often meet the California Specification 01350 standard. As a result, Specification 01350, while not legislated beyond California, has become a nationwide *de facto* regulation [11, 12].

10.6 Codes

A code is a system of rules. For example, the US Code is the compilation of all federal laws. A building code is a set of rules or standards by which a building is constructed. These rules ensure quality, safety, accessibility, and a limited set of environmental issues including energy performance. A building code's rules, or standards, become a law when a municipality, city, or state enacts or votes to make them legally binding. This is referred to as adopting a code.

The International Building Code (IBC) regulates building construction across the United States, including egress, accessibility, and structural requirements. To establish rules for construction, the IBC references standards set by national standards setting organizations (see Table 3.4). States, municipalities, or cities vote on whether the entire IBC is required or just portions. In some cases, states may require local codes by law that are more stringent than the IBC. For instance, in California, buildings must meet seismic codes.

To lessen environmental impacts within the building industry, the International Green Construction Code (IgCC) was established. IgCC rules and standards provide direction on:

- Water and energy efficiency
- Indoor environmental quality
- Site sustainability

- Materials and resources including components, equipment, and systems
- Building commissioning, operations, and maintenance

The IgCC was developed as a model code. Currently, it is rarely implemented as law. Its adoption is through reference by voluntary sustainability standards and certifications, including Leadership in Energy and Environmental Design (LEED). Additionally, governmental commissioning and purchasing programs sometimes refer to IgCC requirements within their design specifications.

International Green Construction Code

The IgCC was developed by the American Institute of Architects (AIA, the US professional organization for architects), ASTM International (a standards setting organization), and the International Code Council (ICC) organization in 2010. The 2018 version of the code incorporates the 2030 carbon neutrality challenge of the AIA, with the consensus-based standards setting processes of ASTM International. It also references the technical information of the American Society of Heating, Refrigerating and Air-Conditioning Engineers (ASHRAE) Standard 189.1 Standard for the Design of High-Performance Green Buildings Except Low-Rise Residential Buildings. The US Green Building Council (USGBC) and the Illuminating Engineering Society (IES) are partners as well.

The IgCC is designed to be coordinated and harmonized for application and adoption:

- LEED certification coordinates with many parts of IgCC.
- The code is coordinated to fulfill tax credits and federal rebates.
- Compliance leads to federal, state, and local incentives.
- It aligns with ICC building commissioning standards.
- It follows the principles of the Office of Management and Budget (OMB) Circular A-119, which governs the federal government's employment of private-sector standards.

While regional governments have not voted the IgCC into law, the aspiration is that harmonization with LEED certification and federal agency purchasing policies will result in the IgCC initially becoming *de facto* regulation and may lead to adoption by municipalities and states as required regulation in the future [13].

10.7 Standards and Certifications

Sustainability standards and certifications provide a path toward adopting environmental initiatives that is different from the path of laws and regulations. Sustainability standards and certifications rely upon free market competition and are voluntary. To ensure competition is fair and comparisons between materials can be made, these standards are established to evaluate environmental initiatives. A standard is a rule or set of rules. For example, a standard can define:

- The composition of a material
- The way in which a material is produced
- How a material should be tested

Sustainability standards for building materials and products set criteria or rules across environmental attributes:

- Resource use
- Energy
- Water
- Emissions
- Toxicity
- Social accountability

There is no established rating system that evaluates whether one environmental attribute is more important than another – while carbon is increasingly of most concern, it does not negate the importance of clean water, or properly treating workers. Each standard has a unique distribution of criteria or rules across environmental impacts. A standard makes value judgments based on the distribution and stringency of individual criteria. One standard may heavily weight social accountability, while another may focus entirely on energy use, and yet another may equally distribute criteria across all environmental impacts.

Manufacturers certify their building materials and products to a standard. In order to achieve certification, a building material or product must meet the requirements of a standard. Meeting requirements is verified through testing and documentation. Once those requirements are achieved, then the manufacturer can attach the associated certification mark or ecolabel to the product, the packaging, or some other relevant documentation.

10.7.1 Types of Standards

A standard provides a set of criteria or rules by which a product can be evaluated. There are two primary ways to categorize these rules:

- What is being evaluated?

  - One kind of product, or many types of products?
  - One single type of criteria or many types of criteria?
- How is the building material or product evaluated in relation to the standard?
  - To be certified, does a manufacturer have to fulfill every criterion, or does the manufacturer get some choices?
  - Are the criteria evaluated in a qualitative way by establishing goals, in a quantitative way focusing upon calculations, or somewhere in between?

10.7.2 What Is Being Evaluated?

Evaluating only one kind of product means that the standard is single sector, whereas evaluating many kinds of products means that the standard is multi-sector. Similarly, considering one type of criteria means that the standard evaluates a single-type attribute, and considering multiple criteria means the standard is multi-attribute (Table 10.1).

10.7.3 How Is the Building Material or Product Evaluated in Relation to the Standard?

To achieve certification, a manufacturer's material or product will be evaluated in relation to a standard. This evaluation will check to see if the data for the product fulfills the criteria requirements of the standard. The evaluation process can follow several options (Table 10.2):

- All prerequisite, single level

Table 10.1 Types of building material and product sustainability standards

Sector	*Attribute*		
Type of Product	Type of Criteria	Example Standard	Description
Single	Multi	NSF/ANSI 140	Evaluates a range of environmental impact attributes specific to only commercial carpet products and their manufacturing processes
Multi	Single	GREENGUARD	Certification emphasizes only indoor air quality, providing a standard for low-emission thresholds for formaldehyde, volatile organic compounds (VOCs), and other interior air contaminants. Products certified to this standard range from countertops and flooring to cabinets and furniture to adhesives
Multi	Multi	Cradle to Cradle	A range of building materials and products are certified to the Cradle to Cradle (C2C) Certified Product Standard across environmental impacts such as chemical composition, material reuse, energy consumption, toxicity, and corporate social responsibility

Table 10.2 Certifying to a standard

Requirements for Criteria	Level(s)	Description	Positives and Negatives	Example
All Prerequisite	Single	All criteria are required to be fulfilled, and therefore every criterion is a prerequisite to certification	These standards have the least flexibility but are most transparent. The standard often provides a public, free list of required criteria	ANSI/BIFMA e3 (low emissions, furniture)
All Prerequisite	Multi-level	The standard offers several levels of certification. All criteria are required to be fulfilled at the specific level of certification and the less stringent levels below	More flexible, manufacturers choose at what level they want to certify their product. Incentivizes manufacturers to attain higher levels of certification over time	Cradle to Cradle (many products, many environmental impacts) Levels: basic, bronze, silver, gold, platinum
Prerequisites and Credits (or points)	Multi-level	Some criteria are required as prerequisites. For other criteria, manufacturers can choose which to fulfill, earning credits or points toward certification. With more credits or points earned, the product attains an increasingly stringent level of certification	Most flexible. Manufacturers choose at what level they want to certify their product and which criteria to fulfill to earn credits. The criteria individual manufacturers use to attain certification are not published. This allows protection of trade secrets and tailoring certification to existing practices. It decreases transparency because the specifics of certification are not published	NSF/ANSI 140 (carpet standard, many environmental impacts) Levels: silver, gold, platinum

- All prerequisite, multi-level
- Prerequisite and credit, multi-level

A prerequisite-based standard requires that all criteria must be met. For example, a prerequisite criterion may dictate that "A manufacturer must document that PBTs (Persistent bioaccumulative toxic substances) are not present at 0.1% or greater in the product," indicating that every product certified to that standard has no PBTs present above 0.1%.

The level of stringency of a prerequisite standard is established by the standard since every prerequisite must be met. From one perspective, criteria cannot be so stringent that manufacturers will not participate. On the other hand, an all prerequisite standard can set the bar so high that very few manufacturers participate, but the standard itself becomes recognized as an ultimate benchmark for sustainability.

By contrast, a prerequisite standard with levels or a prerequisite and credit-based standard introduces competition and motivation for manufacturers. Manufacturers may enter at a basic level but then strive for more stringent levels of compliance. This allows a standard to serve a range of manufacturers and their products, achieving basic, silver, or gold level within a single standard.

In a credit-based standard, a certain number of credits must be fulfilled to receive certification. This provides options and allows the manufacturer to tailor the way they earn certification in relation to their manufacturing processes and interests. For example, in a credit-based standard, a criterion may state "A manufacturer will receive one point for documenting that PBTs are not present at 0.1% or greater in the product." Fulfilling this criterion might lead to 1 point toward certification. However, the same credit-based standard may also provide 1 point if a manufacturer reduces energy use by 3%. From the outside, consumers, buyers, and specifiers only know how many points the product earned, or what level of certification the product achieved. The flexibility of the credit system incentivizes participation, but it also masks transparency. It is impossible to know which criteria a product fulfilled simply based on the number of credits earned.

For credit-based standards, some criteria may be worth more than 1 point. In some cases, point value may be based on the assumed cost or labor intensity, rather than the positive effect of the criterion on preserving the environment.

Currently, there is no universally agreed-upon valuation of environmental impacts. When a standard setting organization imposes a hierarchy through criteria selection or point valuation, it is solely the standard setting organization's evaluation of importance. Different organizations can have vastly different priorities. The distribution of criteria across a standard reveals the prioritized environmental impacts. Within a multi-level or credit-based standard, the manufacturer can then pick and choose which criteria they fulfill. This means the manufacturer can further narrow the focus of the standard through the certification process.

While different types of standards may have the same goal, the way they are structured may result in a slightly different path to meet the goal. A prerequisite standard may ban a list of pesticides and chemicals used in feedstock for a material. By contrast, in a credit-based standard, additional points may be awarded as a manufacturer increases the percentage of biobased or organic feedstock. The methods are different, and the stringency varies, but in both cases the goal is the same.

10.7.4 How Is the Measurement of Environmental Impacts Achieved?

There are many ways to measure environmental impacts, but the framework for measurement falls into four categories (Table 10.3):

- Quantitative criteria

- Relative quantitative criteria
- Qualitative criteria
- Disclosure criteria

It is important to understand that dividing criteria into these categories do not evaluate the specifics of what is being measured. A relative quantitative criterion asking for a 20% decrease from a benchmark may be more stringent than a specific quantitative measurement. However, in general, specific quantitative criteria are less flexible and more stringent than relative quantitative criteria.

Multi-sector standards may rely more heavily on relative quantitative and qualitative measurements. This is because it is difficult to set strict quantitative limits that are reasonable across every product sector. By contrast, a single-sector product standard, such as one for carpet, may more easily restrict ingredients, energy use, and water use due to a more limited and consistent range of products and greater consistency in manufacturing processes.

Whether quantitative, qualitative, or something in between, the actual wording of a criterion adds to the complexity of certifying a product to a standard. The following criterion seems quite simple in the definition:

Product **A** may not use more than **X** kWh energy per unit.

On its surface, this seems to be an easy calculation. However, actually completing the calculation may be complex. The following questions quickly arise:

- How is a unit defined? For instance, if product A is carpet, is a unit determined in terms of square feet, or per roll of carpet? Does the unit include the carpet backing?
- Which processes are included in the total sum of energy per unit? Should the manufacturer count the energy at all of their facilities, or only the final assembly?
- Should the manufacturer include the energy used by their suppliers too?

Standardizing the interpretation of individual criteria can occur either through careful writing of the standard or the consistent interpretation of the standard by certifying organizations. The work of these groups is important to creating harmonization across the certification process.

10.8 Standard Development and the Certification Process

To begin, an organization or group must decide that there is a need for a standard. This group is the Standards Developing Originating Organization (SDOO). Depending upon their expertise with standards, the group may write the standard, or they may contract with a Standards Developing Organization (SDO) in order to develop the standard (Fig. 10.4). In most cases, the goal is to create a standard that serves the sector and becomes regularly employed. In some cases, an organization writes an extremely stringent standard – where it is expected that few products will

Table 10.3 Organizing criteria into categories

Criteria Category	Description	Examples
Quantitative Criteria	Set specific numerical benchmark, limit, or a ban on a process, feedstock, or material Quantitative criteria can be calculated, and products can be compared	Document that PBTs are not present at 0.1% or greater in the product State all energy is documented as renewable Ensure that no waste is discharged into adjacent bodies of water
Relative Quantitative Criteria	Reduction in use or increase in efficiency or specific change, often over a period of time Data can be calculated, but calculations are often in terms of current usage. For example, a 3% decrease in energy use over a 3-year term, where no concrete numbers for the actual energy usage are disclosed This type of calculation makes it impossible to compare different manufacturers' products	Require a decrease in the use of material ingredients Require an increase in the percentage of recycled material Ensure energy or water conservancy Require a change in material formulation relative to an established baseline or benchmark
Qualitative Criteria	Establish policies or goals concerning an issue, often involve public statements on change, and result in the least measurable criteria While there may be benchmarks put in place, or bans agreed to in the future, the current manufacturing process is not evaluated through calculation	Set goals and develop strategies for landfill diversion Develop an energy policy indicating a plan for improved efficiency within the manufacturing process of the certified product
Disclosure Criteria	Require that a manufacturer assess a process or inventory and disclose the resulting data through documentation Disclosure can be solely to the standards certification organization or to the public Disclosure can be one of the most difficult criteria to fulfill: manufacturers must consider whether revealing information exposes trade secrets and proprietary formulations or processes. If a standard requires disclosure across the life cycle, manufacturers may find it difficult to gather their suppliers' data	Inventory product parts, characterizing by type or quality of content, such as biodegradable, biobased, post-consumer recycled content Disclose the product formula characterizing according to toxicity Audit energy including the type of fuel source and kwH usage at each stage of manufacturing Establish a water processing audit for each facility where a finished product is assembled or manufactured

be certified but the standard itself becomes an ultimate benchmark against which other standards are measured.

Once the standard is established, manufacturers decide to certify their materials or products to the standard. The manufacturers typically hire a Conformity Assessment Body (CAB) to collect data from the manufacturer and verify that the criteria of the standard are met. When that assessment is complete, and the product is verified as meeting the standard, then the manufacturer is allowed to add the certification mark or ecolabel to the product, its packaging, or its technical documentation. In addition to CABs, there are also consultants and other professionals that help during the certification process.

10.8.1 Standard Developing Originating Organization

A trade association or non-governmental organization (NGO) often conceives of a standard. They are referred to as the Standard Developing Originating Organization (SDOO). A trade association may seek to standardize an element of its product sector, whereas a NGO is often a non-profit that seeks to advance some cause. The trade association or NGO may have previous experience developing testing methods or industry performance standards.

In some cases, the SDOO acts as the standard developing organization (SDO), writing a proprietary standard that is uniquely authored and owned by the organization. A proprietary standard does not necessarily follow accepted consensus-based processes.

In most cases, the originating organization contracts with an established SDO to write the standard. The benefit of working with a SDO is that standardized and

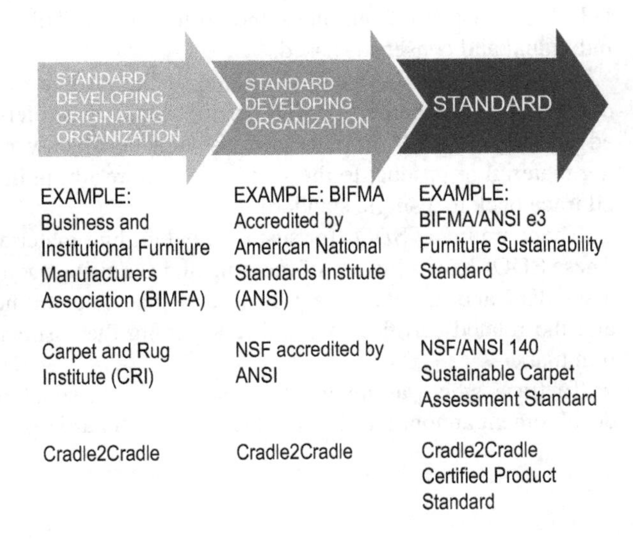

Fig. 10.4 Standards developing structure

accepted methods of procedures, calculations, processes, and testing are applied to the specific subject matter that the trade organization or NGO represents.

10.8.2 Standards Developing Organization

A Standards Developing Organization (SDO) manages the development process of a standard. They ensure that the development process conforms to accepted methods and are often accredited for their expertise. SDOs must balance:

- Stringency to protect the environment in order to satisfy buyers, consumers, and environmental advocates
- Flexibility and ease of implementation to encourage manufacturers to certify their building material or product to the standard

While buyers and consumers are often most familiar with the particular certification mark or ecolabel displayed on a product, the underlying standard establishes the criteria to evaluate the building material or product for its environmental impact.

To accomplish this balance, representatives from manufacturers and buyers form the technical committees that write the standards. This may include authorities from the government, NGO participants, and members from industry. The members of the technical committees agree upon the rules of standards in a consensus-based process, coming to general agreement through discussion. Standards are revisited and changed regularly as new knowledge, technology, and processes become available.

Standards for building materials and products that focus upon environmental impacts are often developed by large, internationally recognized SDOs (Table 10.4).

Each standard is developed entirely separately from all other standards. While some members of technical committees serve on other technical committees, there is little consistency from one standard to another. This is because each process is individual and consensus based.

Sometimes the SDO and the name of the standard are the same. This causes confusion as to whether the SDO or the standard is referenced in publications and advertising materials. In other cases, several certifiers verify compliance of a building material or product to the same standard, resulting in several certifications that all trace back to a single standard.

There are many SDOs for sustainable building materials and product standards. These SDOs have a history of merging and acquiring each other, sometimes leaving a standard and its related certification in place, sometimes renaming the standard and the related certification, and redesigning the associated ecolabel or certification mark.

To trace who was involved in establishing a standard, seek out the standard development annotation. It provides a list of all participants.

Table 10.4 Standards Developing and Accrediting Organizations

Abbreviation	Full Name (in some cases formerly full name)	Area of Focus	Example Sustainability Code	Description
ASTM	American Society for Testing and Materials	Materials, products, services, systems	ASTM E2129-18	Standard practice for data collection for sustainability assessment of building products
ANSI	American National Standards Institute	Accredits SDOs	BIFMA/ANSI e3	Furniture sustainability standard
ISO	The International Organization for Standardization	Management standards focusing on quality, environment, energy	ISO 14044	Environmental management – life cycle assessment – requirements, and guidelines
NSF International	National Sanitary Foundation	Public health and safety	NSF/ANSI 332	Sustainability assessment for resilient floor coverings
UL and UL Environment	Underwriters Laboratory	Public safety and scientific research	UL 100	Standard for sustainability of gypsum board and panels

10.8.3 *ANSI Standards, ANSI-Accredited Standards Developing Organizations, and ANSI-Accredited Certifying Organizations*

Many sustainability standards for building materials and products are ANSI standards. An incentive is that an ANSI-accredited standard provides protection from competition: once a standard is accredited for a particular product sector (category), no other standards in that product sector may become an ANS (American National Standard). However, if a SDO fails to review and update its standard at the end of each 3-year renewal period, then it may be replaced by a standard from another organization, provided the ANSI process is followed.

Accreditation of a standard by ANSI signifies that the procedures used by the SDO meet the Institute's requirements for openness, balance, consensus, and due process in the development of the standard [14]. An ANSI standard may be listed simply by its ANSI name or co-listed by the SDO name and the ANSI name, for example, NSF/ANSI 140.

Within the process, a SDO assembles the joint committee of stakeholders (CS). Stakeholders may include manufacturer representatives, consumers, academics, government officials, and members of non-governmental organizations. Often members of a Standards Developing Committee will serve on another committee of stakeholders, providing expertise from establishing and managing their own standard development process. While there is consistency of process, there is no template that standards follow. Instead, writing a standard relies on the expertise of the participants and the specifics of the sector.

ANSI accredits SDOs and CABs for certification of products. This ensures these organizations can establish standards and provide certification methods that follow ANSI consensus-based processes.

History of ASTM and ANSI

Two of the most widely recognized standards setting organizations in the country are ASTM (American Society for Testing and Materials) and ANSI (American National Standards Institute). Both organizations publish a variety of standards on:

- Testing products
- Providing services
- Establishing processes and procedures
- Overseeing systems
- Managing personnel

ASTM was established in 1898 in response to a need for standardization of material production. Its origin was in specifying steel composition in railroad rails. Initially, the Pennsylvania Railroad issued a specification for the chemical makeup of the steel in railroad rails, but the rail suppliers denied that the required composition was necessary. As conversation between the railroad and the suppliers unfolded, it became clear that different parties offered different information ranging from practicalities of production and cost structure to performance and requirements for the equipment using the rails. By talking through issues, what seemed intractable was solved through consensus.

Charles Dudley, first president of ASTM, stated [15]:

> A good specification needs both the knowledge of the product's behavior during manufacture and knowledge of those who know its behavior while in service.

In 1902 ASTM established specifications for cement and concrete. Both steel and concrete were engineered products. They required specific methods of

(continued)

production and chemical composition to achieve performance. This was unlike naturally sourced materials such as wood and stone, which could be selected by species or type. The development of the concrete specification focused not only on production and composition but also on the methods of testing the material, another opportunity for standardization.

From the beginning, negotiation between supplier and buyer formed the standard's development process. Additionally, participation in development, adherence, and use of a standard was voluntary; and standards were replaced as advancements occurred in material composition, manufacturing processes, and testing procedures. Within a standard's development technical committee, suppliers' representatives could not outnumber buyers' representatives, and a supplier could not head a committee. This structure established checks and balances to ensure fairness of outcome.

The American National Standards Institute (ANSI) was founded in 1916 by the American Institute of Electrical Engineers (IEEE), American Society of Mechanical Engineers (ASME), American Society of Civil Engineers (ASCE), American Institute of Mining, Metallurgical, and Petroleum Engineers (AIME), and ASTM. Its first standard was to develop consistency in pipe threads. Its next set of standards focused upon safety codes. Other standards included mining methods, electrical and mechanical engineering, and traffic safety. After World War II, ANSI became a leader in the International Organization for Standardization (ISO). Today ANSI accredits Standards Developing Organizations to ensure they follow the ANSI process of openness and consensus when they develop standards[16].

10.8.4 Governmental Standards

US governmental agencies sometimes reference standards in relation to legislation or regulations. These are often private standards developed through the consensus process.

> In the United States, government agencies are required to use existing private sector standards wherever feasible as the basis for technical regulations and for government procurement bids. This practice has many benefits, advancing both public interest (efficient protection of environment, health and safety) and economic development [17].

By contrast, EU governmental organizations fulfill the roles of the SDOO, SDO, and CAB.

10.8.5 Conformity Assessment Bodies

Once a standard is developed, a product is evaluated as to whether it meets the standard (Fig. 10.5). The product can be certified through:

- First-party self-certification
- Second-party certification
- Third-party certification

Third-party certification is performed by an independent laboratory or Conformity Assessment Body (CAB). First-, second-, and third-party certification processes are defined by the ISO 14020 standard (Table 10.5).

Whether first-, second-, or third-party, the CAB evaluates and certifies an applicant's product against the standard's criteria. Depending on the specificity of the standard document, the CAB may play a large role in interpreting and determining the stringency of the criteria. As a result, this process can vary in intensity and transparency depending upon the quality of the standard's construction.

10.8.6 Auditors and Consultants

Auditors observe and evaluate the assessment process. This ensures that different CABs or their members interpret a standard consistently and certify products in the same way; some standards also require annual or random field audits to ensure continued compliance to the standard during the manufacturing process.

Manufacturers often hire consultants to help lead them through the certification process. Standards are often complicated, and understanding how best to

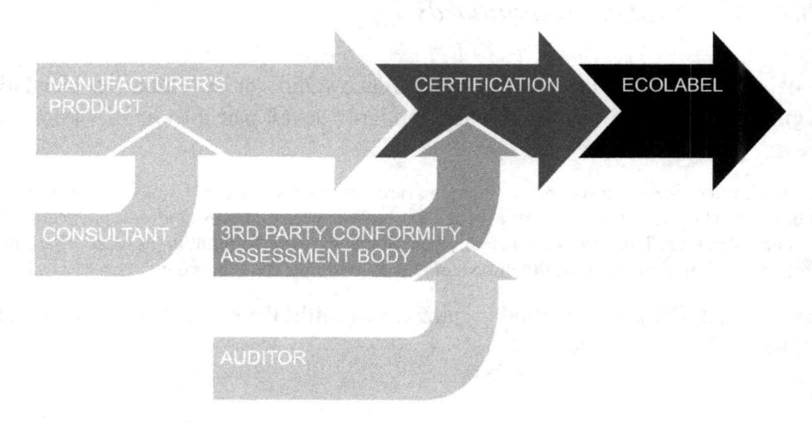

Fig. 10.5 Certifying a manufacturer's product

evaluate and assess a specific product for conformance can be confusing. Sometimes this is because a standard may be very prescriptive, and narrowly focused in a way that does not obviously apply to a product. Conversely, a standard might be vague in an area. Additionally, many standards are point based, so not all criteria must be met.

Deciding which criteria to pursue has implications for a manufacturer in terms of:

- Time to research the current manufacturing process
- Cost to possibly change the manufacturing process
- Selection of raw materials and how that might impact the choice of suppliers
- The amount of research and development required to alter a product to earn a point toward the certification
- Whether changing the manufacturing process or supply chain to earn a credit may affect the product's performance, quality, or design aesthetics

Third-party certification bodies are not permitted to act as consultants to help in a manufacturer's decision-making, because it would present a conflict of interest during the certification process. However, many certification bodies will approve or provide consultants to work with manufacturers in preparation for certification. SDOOs such as trade associations also often provide online tools or educational seminars in preparation for certification.

Table 10.5 Understanding first-, second-, and third-party certifications

Certification Type	Description	Possible Positives	Possible Negatives
First-Party	Self-assessed	May reveal inefficiencies or failures in the manufacturing process	Often this is an internal evaluation Not considered independent or unbiased
Second-Party	Completed by a body that has an interest in the manufacture of the product For example, SDO certifying its own products	Useful as an internal audit Can be a manufacturer evaluating a sub-contractor or supplier	May be biased
Third-Party	Completed by an independent review body or laboratory	Most objective Standards approved by ANSI require third-party certification	Most costly and involved

10.9 Ecolabel or Certification Mark

Once an accredited second- or third-party verifies certification of a building material or product, an ecolabel or certification mark can be affixed to the product or its packaging. A product's documentation or specifications may also include a reference to the certification. It is important to remember the ecolabel is related to the certification, and the CAB, not the standard. Several different CABs could certify to a single standard, with each organization issuing its own organization's ecolabel.

An ecolabel or certification mark contrasts with "green" symbols or claim statements developed by manufacturers and service providers. Green symbols and claim statements such as "healthy," "natural," or "safe" are unsubstantiated declarations. They are essentially marketing terminology that is unregulated.

References

1. Environmental Protection Agency. (2002). Federal Water Pollution Control Act. Sec. 101 (a) (3), P 3. https://www.epa.gov/sites/production/files/2017-08/documents/federal-water-pollution-control-act-508full.pdf. Accessed 8 Aug 2020.
2. Rodgers, Jr., & William, H. (2001). Executive orders and presidential commands: Presidents riding to the rescue of the environment. *21 Journal of Land Resources & Environmental Law*, *13*, 22–24. https://digitalcommons.law.uw.edu/cgi/viewcontent.cgi?article=1248&context=faculty-articles. Accessed 8 Aug 2020.
3. The American Presidency Project, UC Santa Barbara. (2020). *Executive orders*. https://www.presidency.ucsb.edu/statistics/data/executive-orders. Accessed 1 Aug 2020.
4. Clinton, W. (1994). Executive order 12898, federal actions to address environmental justice in minority populations and low-income populations. *Federal Register, 59*(32). https://www.archives.gov/files/federal-register/executive-orders/pdf/128910.pdf. Accessed 1 Aug 2020.
5. Obama, B. (2009). Executive order 13514: Federal leadership in environmental, energy, and economic performance. *Federal Register, 74*(194). https://www.fedcenter.gov/programs/eo13514/. Accessed 1 Aug 2020.
6. Obama, B. (2015). Executive order 13693: Planning for federal sustainability in the next decade. *Federal Register, 80*(57). https://www.fedcenter.gov/programs/eo13693/. Accessed 6 Aug 2020.
7. Trump, D. (2018). EO 13834: Efficient federal operations. *Federal Register, 83*(910). https://www.fedcenter.gov/programs/eo13834/. Accessed 6 Aug 2020.
8. Supreme Court of the United States. (2006). *Massachusetts et al. v. Environmental Protection Agency et al.* https://www.supremecourt.gov/opinions/06pdf/05-1120.pdf. Accessed 1 Aug 2020.
9. Morrissey, D. (2018). *3 environmental lawsuits that have helped society*. Sustainable Life, Inc. https://sustainablebrands.com/read/leadership/3-environmental-lawsuits-that-have-helped-society. Accessed 1 Aug 2020.
10. Greenhouse, L. (2007). Justices say E.P.A. has power to act on harmful gases. *The New York Times*. https://www.nytimes.com/2007/04/03/washington/03scotus.html. Accessed 1 Aug 2020.
11. Theirault, E. (2018). "Brand Memo" prohibits US DOJ from converting agency guidance into binding legal obligations in civil enforcement actions. *The National Law Review*. https://www.natlawreview.com/article/brand-memo-prohibits-us-doj-converting-agency-guidance-binding-legal-obligations. Accessed 1 Aug 2020.

12. CalRecycle. (2019). *Section 01350: Special environmental requirements*. https://www.calrecycle.ca.gov/greenbuilding/specs/section01350. Accessed 1 Aug 2020.
13. International Code Council. (2018). *Overview of the international green construction code*. https://www.iccsafe.org/products-and-services/i-codes/2018-i-codes/igcc/. Accessed 1 Aug 2020.
14. American National Standards Institute. (2020). *ANSI essential requirements: Due process requirements for American National Standards*. https://share.ansi.org/Shared%20Documents/Standards%20Activities/American%20National%20Standards/Procedures,%20Guides,%20and%20Forms/2020_ANSI_Essential_Requirements.pdf. Accessed 30 July 2020.
15. ASTM International. (2017). *The history of ASTM international*. https://www.astm.org/ABOUT/history_book.html. Accessed 29 July 2020.
16. ANSI. (2018). *100 years of ANSI*. https://www.ansi.org/about_ansi/introduction/history. Accessed 29 July 2020.
17. American National Standards Institute. (2020). *U.S. standards system: Government use of standards*. https://standardsportal.org/USa_en/standards_system/government_use_standards.aspx. Accessed 1 Aug 2020.

Chapter 11
Transparency

11.1 Introduction

The certifications, standards, and ecolabels presented in Chap. 10 are a way to demonstrate a manufacturer's sustainability initiatives. Adding a symbol or name allows a manufacturer to communicate they are making a deliberate effort to be sustainable with their building materials and products. However, certifications and ecolabels are often awarded based upon scorecards that are unreleased to the public. Specifiers, owners, designers, and non-governmental organizations (NGOs) have become more educated on the characteristics of sustainable materials during the last several decades. As a result, they now ask for transparency on how manufacturers specifically achieved certification – how they fulfilled the certification scorecards. This signals a change. Instead of goals focused upon specifying products that are sustainably certified, the design profession now looks for specific criteria and transparent disclosure of data.

Why Transparency?

An architect receives a commission to design a new headquarters. The client announces a healthy indoor air quality (IAQ) program as part of their sustainability goals related to employee health and wellbeing. The client asks the architect to ensure all material finishes on the project meet the indoor air quality standard the client sets for low volatile organic compound (VOC) content and emissions. The architect decides to use certifications or ecolabels to select products that meet the low VOC content.

First, the architect finds an ecolabel that advertises IAQ on its website and scorecard. The architect identifies the standard backing the certification to uncover what requirements are included and whether they meet the client's VOC standards.

(continued)

H. R. Roth et al., *The Green Building Materials Manual*,
https://doi.org/10.1007/978-3-030-64888-6_11

Next, the architect must determine whether the VOC requirement is a prerequisite or a credit. If VOC limits are a prerequisite and the limits meet the client's standards, then any product with that label can be used. However, if VOC limits are a credit (i.e., voluntary) requirement, only a portion of certified products will use this credit to score points. There is no guarantee that a specific product meets the VOC limits credit. Only if the certification publishes each product's scorecard can the architect be assured that the VOC limit is met.

Instead, the architect may select an ecolabel that focuses only on VOC requirements, like GREENGUARD. However, if the client has other requirements, the architect may need to find products that earn several certifications, each satisfying specific criteria. Identifying products with multiple certifications reduces the number of finishes available for selection.

Company-wide or client-led goals and requirements related to tangible environmental, social, and health indicators have increased. Clients have specific goals that they want to meet that go beyond certification. This specificity has encouraged more straightforward paths to identifying a material's ingredients and environmental impacts. Two solutions in particular have emerged: disclosure and product databases.

11.2 Disclosure

This chapter focuses on the disclosure of sustainability-related data from a manufacturer directly to the public on a specific product. Disclosure can include:

- List of ingredients
- Energy consumption amounts
- Carbon footprint per unit of product
- Publication of policies related to labor and human rights for a particular factory

Disclosure can be a difficult hurdle for building material manufacturers. It often requires the collection of data across a complex supply chain, which may span multiple countries, tens or hundreds of facilities, and hundreds or thousands of individuals. It takes both cost and time to collect data. When contracts change with suppliers, new data needs to be collected. Additionally, manufacturers or their supply chain partners may be reluctant to reveal certain data if it is connected to a trade secret unique to their product.

To date, the building and construction industries have not yet agreed upon a range of data that must be published regarding a building material. This idea is often referred to as a "nutrition label for building materials," referring to the FDA nutrition facts label required for food products. In addition to the complexity of supply chains, it is difficult to identify a concise list of attributes for a label that is relevant across different product sectors. Each sector has different institutional history, methods of production, and geographic location. Examples of variations in measurement and reporting of data include:

- Methods of measurement: Can impacts be measured in the same way across sectors? Do standards exist that relate?
- Precision of measurement: At what threshold of content is a material identified as an ingredient? More than 1% of the product by weight? More than 0.1%? More than 0.001%?
- Omissions: Are byproducts included? Are the impacts of recycling an old material into new considered, or only impacts once a recycled feedstock arrives at the factory?
- Issues of comparability related to life cycle assessments (see Chap. 3).

There are two types of declarations that attempt to take on the role of a nutrition label for building materials. Environmental Product Declarations (EPDs) disclose data in terms of emissions and environmental impacts. Health disclosures provide ingredient lists and their related toxicity and health impacts. While a consistent framework for disclosure is in place, in some cases competing rules exist or have not yet been agreed upon.

11.3 Material Inventories and Characterization

The list of ingredients that makes up a product is a bill of materials or a material inventory. The term "material inventory" originates in business: a raw material inventory is the cost of all materials or components in stock that are not contained in work-in-progress or finished goods. A manufacturer asks for a material inventory from its suppliers. If a manufacturer cannot get a full list of ingredients, it may choose to test a material to get its material characterization. Material characterization is a term from materials science, referring to the probing and measuring of a material's structure, physical properties, and chemical makeup.

The accuracy of the measurement of ingredients and chemical makeup varies depending on the intended goal of the material inventory. A bill of materials intended for use as a list of inputs for a life cycle assessment may include only ingredients that make up greater than 1% of the total mass of the product. A bill of materials intended to evaluate toxicity must be more specific, reporting ingredients at the threshold of 0.01% (100 ppm) or 0.1% (1000 ppm). This difference in specificity and reporting thresholds is primarily due to two issues: impact threshold and certainty (Table 11.1).

Table 11.1 Impacts and certainty of measurement

Issue	Description	Example
Impact Threshold	Very small percentages of chemical ingredients can be toxic enough to impact human health and so should be disclosed in a Health Product Declaration (HPD) Very small percentages of chemical ingredients are not likely to impact overall emissions calculations, and as a result, do not need to be reported in an Environmental Product Declaration (EPD)	A very small amount of formaldehyde in a product has been proven to be hazardous to human health and should be reported By contrast, nails and other fasteners are often excluded from a whole building life cycle assessment, because they are such a small percentage of the weight of an entire building that they will not be identified in the top 10 (or 50) material hot spots (emissions contributors) for a building
Certainty	Characterizing ingredients for negative impact on human health has a high degree of certainty: Federal testing establishes thresholds for toxicity Characterizing ingredients for their life cycle impacts is less certain as many factors determine those impacts	Scientists have established the amount of formaldehyde in a glue that can cause negative human health impacts It is much more difficult to determine how many megajoules (MJ) of energy are required for the extraction and processing of the same glue – evaluation would be based on the average amount of energy required across a variety of fuel sources, geographic location of extraction and production, and the type of manufacturing equipment used Additionally, the LCA practitioner may make assumptions about which stages of LCA are included in the evaluation To state that the glue required 11.378 MJ of energy per gram of glue would be misleading, as it would imply a level of certainty that the practitioner is not able to provide

11.4 Environmental Product Declarations

An Environmental Product Declaration (EPD) is a voluntary declaration of data. The data is evaluated based on the framework of life cycle assessment (LCA) described in Chap. 3. Unlike most certifications and ecolabels, EPDs do not rank materials or products for meeting a certain level of environmental performance criteria: there is no silver, gold, or platinum EPD. EPDs are disclosure documents that allow manufacturers and consumers to make decisions according to their own values and priorities, using the information disclosed about the processes and resources being used, and the potential environmental impacts of the material or product. Similar to a food nutrition label, if the consumer does not know what a "good" amount of carbohydrates or fats or protein is, then the label will not be useful for making decisions.

EPDs are standardized. They include the following information:

- Reference to the Product Category Rules (PCRs) used for the EPD
- Declaration of unit size for calculation
- Declaration of systems boundary
- Disclosure of assumptions, estimates, cut-off criteria, data quality, and period of review
- Description of the system boundaries of study (see life cycle stages above)
- Characterization method (i.e., TRACI 2.1 (2012), etc.)
- Summary of life cycle impact assessment, typically including global warming potential (GWP), stratospheric ozone layer depletion potential (ODP), acidification potential (AP), eutrophication potential (EP), formation potential of tropospheric ozone, photochemical ozone creation potential, and depletion potential for fossil fuel and non-fossil fuel resources

Because EPDs are based on life cycle assessments, they may include resource, energy, and water use information that was used to calculate the life cycle impacts of the product, such as:

- Use of primary and secondary materials
- Use of freshwater
- Use of renewable and non-renewable primary energy, including transportation
- Use of renewable and non-renewable secondary fuels
- Output flows and waste
  - Hazardous, non-hazardous, radioactive waste disposed
  - Components for reuse or recycling
  - Energy recovery from material waste
  - Exported energy

EPDs may also include general sustainability and manufacturing data for a product, such as:

- A product description in relation to CSI MasterFormat and its related standard specification
- List of ASTM standards governing type, grade class and use, and product
- Material composition, based on Safety Data Sheets (SDS) Rules
- Manufacturing processes and locations
- Programs specific to the product that go beyond legislation or guidelines with respect to emissions, water discharge, and noise
- Disclosure of industrial and environmental protection during product processing and/or installation
- Processing installation, including the description of machinery, tools, and dust extraction equipment
- Description of packaging including type, composition, ability to reuse or recycle
- Disposal methods including reuse/recycling/energy recovery/other required disposal methods and taking into account hazards
- Performance when exposed to fire, water, or destruction

The creation of EPDs is guided by several international standards, including ISO 14025 and EN15804, and ISO 14044. Product Category Rules (PCRs), described later in this chapter, provide a framework for standardizing what data is disclosed and how it is evaluated within a specific product sector, such as "Flooring" or "Non-Metal Ceiling Panels" [1].

Third-party published EPDs are regarded as independent and without bias. These EPDs are called Type III and are verified by a third-party assessor and overseen by a program operator ensuring compliance with ISO 14025 guidelines. The American Center for Life Cycle Assessment (ACLCA) maintains a list of EPD program operators [2]. They include:

- SCS Global Services (SCS)
- Underwriters Laboratory Environment (ULE)
- Sustainable Minds

While EPDs are necessary and valuable tools for providing product-specific life cycle impact data, they do have drawbacks. EPDs are currently expensive to create and require ongoing investment, as they are only valid for 5 years. Small-scale companies may not be able to invest in the required evaluation and testing of data, and it is therefore less likely for them to seek an EPD. Additionally, certain materials and product sectors have few or no completed EPDs. As a result, new EPDs in those sectors do not have a history of data to rely upon for comparison. This makes developing an EPD more time-consuming and labor-intensive. It can also introduce more assumptions and variations in the scope of evaluation.

There are three primary strategies for increased adoption of EPDs: legislation requiring adoption, certification incentivizing adoption, and free databases increasing access to data (Table 11.2).

11.4.1 Product Category Rules

Product Category Rules (PCRs) provide a framework for EPDs across a specific material or product sectors such as "flooring" or "non-metal ceiling panels." This framework specifies what input and output data should be included for all EPDs within the sector. PCRs provide:

- Consistent scope, system boundary, and measurement procedures.
- Calculation methods for tabulating impacts and generic data for calculations where data is difficult to collect.
- A list of companies that provided data to establish the PCR. Typically these companies have developed EPDs for their products within the sector.
- Additional data requirements outside the scope of an LCA, such as disclosure on human health and toxicity (if determined as relevant by the PCR Committee).

Historically an array of PCRs exists within single sectors. The International Organization for Standardization (ISO) currently seeks to harmonize PCRs to

Table 11.2 Strategies for increasing adoption of EPDs

Method	Description	Example
Voluntary Standards and Certifications	Sustainable whole building certifications provide points toward certification to incentivize publishing an EPD	LEED v4 provides points toward certification for the use of 20 or more different products (from at least 5 manufacturers) with an LCA or EPD [3]
Required State Legislation and Regulation	Procurement program used to create *de facto* regulation requiring EPDs	Buy Clean California Act, passed in October 2017, requires products for public projects have an EPD Materials, such as structural steel, rebar, flat glass, and mineral wool board insulation, must be evaluated for global warming potential (GWP) through EPD Data collected by California from these EPDs will establish GWP levels for each product After July 1, 2021, eligible construction materials in California must be at or below the established GWP level for that material, as verified by EPD
Free EPD Databases	Provide industry professionals with a tool to find and compare EPDs	Embodied Carbon in Construction Calculator (EC3) UL Spot

remove redundancy and to promote comparison between individual EPDs. ISO seeks to establish one PCR for each product sector, simplifying disclosure while preserving variation due to geographic location and particular manufacturing or processing methods. Once a single PCR is established for a sector, ISO 14025 requires that all individual EPDs be prepared in alignment with the criteria of that PCR.

The process for developing PCRs is similar to the development of other standards. An open development process is required with the engagement of a range of stakeholders and review methods similar to those presented in Chap. 10. Stakeholders typically include product manufacturers and industry associations for the relevant sector. An independent agency called a program operator must supervise the entire PCR development process and ensure that ISO 14025 is followed. PCRs must be updated every 3–5 years to stay valid.

11.5 Health Disclosures

Health Product Declarations and the Declare Label both require disclosure of material ingredients and then evaluate those ingredients for toxicity and human health impacts using established chemical red lists.

11.5.1 Health Product Declarations

A Health Product Declaration (HPD) provides information on the chemical and mineral composition of a material or product and associated health risk. Key components of an HPD include (Table 11.3):

- Content inventory: a list of material ingredients, based upon the threshold of measurement, 100 ppm, 1000 ppm, or as defined by another body (such as a SDS). These may or may not include materials that are considered residuals or impurities.
- Composition of ingredients tabulated by the percentage of weight, which can include a range of percentages (i.e., the composition of a product including "X" chemical which may range from 0.1% to 0.5% dependent on the specific product).
- Hazard assessment of each ingredient, as determined by screening against hazard lists, including whether those hazard ingredients actually represent a legitimate risk based on installation and/or use. The HPD Priority Hazard List is organized based upon (Table 11.3):
  - GreenScreen Benchmark Score – with benchmarks defining progressively safer materials

Table 11.3 Understanding GreenScreen Benchmark and GreenScreen List Translator Score

GreenScreen Benchmark Score	Scores Based on a GreenScreen Assessment Approach
Benchmark 1, Chemical of High Concern	Chemicals of high concern that should be avoided, including carcinogens; reproductive, developmental, and neurodevelopmental toxicants; mutagens; persistent, bioaccumulative and toxic chemicals (PBTs); very persistent and very bioaccumulative chemicals (vPvBs); and endocrine disruptors
Benchmark 2	Use but look for safer substitutes
Benchmark 3	Use but has the opportunity to be improved
Benchmark 4	Preferred chemical – considered safe
GreenScreen List Translator Score	**Evaluates against 40 hazard lists from scientific bodies, NGOs, and government agencies** [4]
LT-1	Chemical meets one or more of the GreenScreen Benchmark 1 criteria and would likely be identified as a Benchmark 1 chemical if a full GreenScreen assessment were conducted
LT-P1	Chemical meets one or more of the GreenScreen Benchmark 1 criteria Greater uncertainty about toxicity Research is needed to determine whether it would likely be identified as a Benchmark 1
LT-UNK	This chemical was present on a chemical of concern list Insufficient information to evaluate the specific toxicity of the chemical in this situation

  - GreenScreen List Translator Score – chemicals evaluated against red lists and chemicals of concern lists provided by scientific bodies, NGOs, and governmental agencies
- Whether nano-materials are utilized.
- Volatile organic compound (VOC) content.
- A list of certification and ecolabels associated with the product and whether the product works toward earning credits toward Leadership in Energy and Environmental Design (LEED) certification.
- Whether the HPD is third-party verified and who prepared the HPD (self-prepared, in the case of a non-third-party-verified HPD).
- Publication and expiration date.

The 40 hazard lists utilized by the GreenScreen Assessment List Translator were developed by agencies such as:

- The Environmental Protection Agency (EPA) and the Centers for Disease Control and Prevention (CDC)
- European Union/European Commission
- International Agency for Research on Cancer, World Health Organization
- Intergovernmental Panel on Climate Change
- Directives from Japan, Korea, Germany, Malaysia, New Zealand, Australia
- State legislation from Oregon, California, and Washington

To preserve proprietary formulations, all ingredients are screened against HPD hazard lists, but not all CAS numbers are disclosed. Manufacturers can also disclose a range of percentage makeup rather than a specific number to account for variation in manufacturing.

The declared data of an EPD references the EPA's TRACI 2.1 format, SDS, and a specific list of information based on PCRs. By contrast, HPDs evaluate chemical hazards by referencing directives, regulations, and laws ranging from international treaties to state mandates.

HPDs are available for an increasing number of products. A number of green building standards incentivize manufacturers to create HPDs by requiring or rewarding points for projects that have HPDs, including Cradle to Cradle, BIFMA, LEED, and WELL. For example, LEED v4 awards points for the Building Product Disclosure and Optimization – Materials Ingredients credit for projects that specify 20 or more materials with HPD (or a range of other disclosures) sourced from a minimum of 5 manufacturers.

The Health Product Declaration (HPD) Collaborative is a NGO. It is membership based. Led by the Healthy Building Network (HBN) and BuildingGreen, the collaborative established the HPD Open Standard in 2012. It is a voluntary, consensus-based standard. HPD is a proprietary standard that is not accredited by ISO, ASTM, or ANSI. Manufacturers can elect to self-report material composition or to use third-party verification.

Table 11.4 Understanding ILFI's Declare Label

Category	Description
Red List Free	Disclose 100% of their product's ingredients and residuals present at or above 100 ppm (0.01%) Do not contain any red list chemicals at 100 ppm
Red List Approved	Disclosed 99% or more of their product's ingredients and residuals present at or above 100 ppm (0.01%) Products meet the requirements of the Living Building Challenge Red List Imperative Products may contain one or more red list chemicals if the ingredient falls under an exception
Declared	Disclosed 100% of their product's ingredients and residuals present at or above 100 ppm (0.01%) Contain one or more red list chemicals

Table 11.5 Product- and brand-specific resources

Product Database	Number of Products	Specialization	Organization
ULE Spot	130,000 product families	General/holistic	UL
mindful MATERIALS	20,000 materials	General/holistic	GIGA
Sustainable Minds Transparency Catalog	1500 brands represented	General/holistic	Sustainable Minds
GreenSpec	80 product categories, each with 1+ products	General/holistic	GreenSpec (UK)
Embodied Carbon in Construction Calculator (EC3)	33,000+ products across 7 categories	Embodied carbon	Building Transparency
Health Product Declaration (HPD) Public Repository	6800+ products	Health	HPD Collaborative
SF Approved Product List	300 products	Health	SF Environment

Table 11.6 General product or specification guidance resources

Resource	Number of Product Types	Specialization	Organization
Pharos Project	165,476 chemicals 164 building products	Health	Healthy Building Network (HBN)
QuartzProject	102 products	Health and embodied carbon	Google, Flux, HBN, thinkstep (no longer being updated)
BuildingGreen, Product Guides	110 product categories	General/holistic	BuildingGreen
HomeFree	9 product categories	Health	HBN

11.5.2 *The Declare Label*

The Declare Label is another transparency-focused health initiative. It provides a nutrition label-style disclosure of ingredients for a building product. The Declare Label is closely tied to the Red List maintained by the International Living Future Institute (ILFI) used by the Living Building Challenge, Living Product Challenge, and Declare Label [5]. Similar to the HPD database managed by the HPD Collaborative, the International Living Future Institute manages an online database of all products that have a valid Declare Label.

In addition to a list of ingredients, the Declare Label divides products into one of three categories: Red List Free, Red List Approved, or Declared (Table 11.4) [6].

11.6 Data at Your Fingertips: Databases

One key to transparency is easy access and comparison of the sustainability indicators for different building materials and products. A range of databases exists that aim to provide this ease of comparison and research for users. Different databases focus on different subject matter and provide different representation across market share (Tables 11.5 and 11.6).

References

1. ULE. (2020). *Product Category Rules (PCRs)*. https://www.ul.com/offerings/product-category-rules-pcrs. Accessed 12 Sept 2020.
2. The American Center for Life Cycle Assessment (ACLCA). (2020). *Program operators*. https://aclca.org/pcr/program-operators/. Accessed 12 Sept 2020.
3. Lasso, A. (2017). *Decoding environmental product declarations*. Green Business Certification, Inc. https://gbci.org/decoding-environmental-product-declarations. Accessed 12 Sept 2020.
4. GreenScreen for Safer Chemicals. (2020). *GreenScreen List Translator*. https://www.greenscreenchemicals.org/learn/greenscreen-list-translator
5. International Living-Future Institute. (2020). *The red list*. https://living-future.org/declare/declare-about/red-list/. Accessed 12 Sept 2020.
6. International Living-Future Institute. (2020). *What does a product's Declaration status symbolize? FAQ, Declare*. https://living-future.org/contact-us/faq/. Accessed 12 Sept 2020.

Chapter 12
Conclusion

12.1 Introduction

This book presents current environmental initiatives and sustainability metrics related to building materials. This chapter concludes with a snapshot of some of the people and events that led to the development of these ideas and strategies. This selective timeline of historical progress related to sustainability demonstrates how the evolution of building material standards, certifications, legislation, and regulation is a reflection of the changing thinking of society at large.

Systemic change can only result from the sustained collective effort of sizable groups of people. It would be impossible to capture even a fraction of the web of people, actions, and events that led to the issues laid out in this book. Rather than a comprehensive history, this section presents a set of narratives that capture some of the moments and historical patterns that have led to today's current initiatives.

The examples included in this chapter reflect the majority of this book in that they focus on events and legislation in the United States, with occasional reference to examples in Western Europe. There are significant examples of standard development and sustainable building materials initiatives across the world that are not included in this chapter.

12.1.1 Why People?

The history of sustainability is a history of people rather than a history of the planet. It is a difficult one to tell succinctly, as each individual and event is a product of a much broader group of interconnected people, actions, and events. In 2007, Paul Hawken attempted to capture a portion of the vast number of organizations and communities that make up the environmental and social justice movement in his

H. R. Roth et al., *The Green Building Materials Manual*,
https://doi.org/10.1007/978-3-030-64888-6_12

book *Blessed Unrest: How the Largest Movement in the World Came into Being and Why No One Saw It Coming*:

> Across the planet, groups ranging from ad hoc neighborhood associations to well-funded international organizations are confronting issues like the destruction of the environment, the abuses of free-market fundamentalism, social justice, and the loss of indigenous cultures. They share no orthodoxy or unifying ideology; they follow no single charismatic leader; they remain supple enough to coalesce easily into larger networks to achieve their goals. While they are mostly unrecognized by politicians and the media, they are bringing about what may one day be judged the single most profound transformation of human society. [1]

12.2 A Timeline

12.2.1 Pre-1960

From the perspective of 2020, one can easily look backward and recognize how many environmental efforts from the past 60 years have focused on remedying the unintended damage and destruction brought about by earlier actions (Table 12.1). Companies, corporate entities, and governmental entities organized, designed, and expedited industrial processes and infrastructure dating from the Industrial Revolution in the late 1700s to present day. Growth and expansion consistently have been perceived as synonymous with prosperity and an improved standard of living.

Beginning in the ninth century, Viking settlers razed the birch woodlands that covered approximately 25% of Iceland in a period of 300 years, burning the forest for agriculture and grazing land or using it for timber construction and charcoal [2]. As a result, Iceland is now approximately 40% desert, with severe soil erosion and an inability to farm a large part of the country. While the ninth century is not the first irrevocable environmental destruction resulting in our current geography, it provides an example 11 centuries ago of humans overburdening our ecosystems to the point of destruction.

Historical cause and effect are not always a straight line between two and three events: overlapping storylines and trends have unforeseen effects. In 1908, the production of the Model T began. The Model T was the first affordable car, shifting away from a luxury item for the wealthiest Americans to a household item necessary for daily life. The Model T was greeted enthusiastically by the public. Concurrently, iron and steel production increased with industrialization in the United States including the production of materials to satisfy the military needs of World War I and World War II. Polluted industrial cities characterized the US post-World War II. Buildings across the United States were black from burning coal to run industry, whether producing steel, iron, aluminum, autos, or airplanes. As pollution increased, political momentum for environmental action and legislation grew. This heightened awareness eventually resulted in the environmental action and legislation associated with the 1960s.

Concurrently, the foundation of the organizations that today lead global conversations related to climate change and sustainability occurred. Two notable examples

Table 12.1 A brief timeline of events covered in this book: pre-1960

Pre-1960	
1760	The Industrial Revolution began in Britain, spreading across the world from 1760 to about 1840
1853	The British Parliament passed the Metropolitan Smoke Abatement Act to limit the production of smoke in London
1875	The British Parliament enacted the Public Health Act, expanding limits on smoke across all of Britain
1908	The Model T was sold by the Ford Motor Company
1914	World War I began
1929	The Great Depression began
1930	The Civilian Conservation Corps Executive Orders were issued by Franklin Delano Roosevelt to prevent land erosion and provide reforestation after the Dust Bowl (see Chap. 10)
1935	Congress passed the National Labor Relations Act, establishing workers' rights to form unions (see Chap. 9)
1938	Federal Highway Act of 1938 was passed into law, initiating research into the Interstate Highway System [3]
1939	End of the Great Depression in the United States Beginning of World War II
1945	World War II ended The United Nations (UN) was officially founded by representatives of 50 countries at the United Nations Conference on International Organization [4]
1947	The International Organization for Standardization (ISO) was created. The ISO began with 67 technical committees (groups of experts focusing on a specific subject). As of 2020, the ISO has 792 technical committees representing 165 members covering 23,302 technical standards
1948	The Federal Water Pollution Control Act (FWPCA) was established, later replaced by the Clean Water Act (CWA) (see Chap. 6) The United Nations proclaimed the Universal Declaration of Human Rights (UDHR), established international human rights law (see Chap. 9)
1952	The London Fog/Smog event killed approximately 4,000 people in London and sickened another 100,000, resulting in the 1956 British Clean Air Act (see Chap. 7)
1955	*Life* magazine described "Throwaway Living" (see Chap. 4) Congress enacted the first federal air pollution law, the Air Pollution Control Act (see Chap. 7)

include the United Nations, recognized for its treaties on the environment and human rights, and the International Organization for Standardization (ISO), which provides standards for evaluating sustainability and worker protection.

12.2.2 1960–1979

The 1960s saw the civil rights protests; the assassinations of President John F. Kennedy, Dr. Martin Luther King, and US Attorney General Robert Kennedy; the passage of the Civil Rights Law of 1964, the Vietnam War, and the anti-war protests.

The period from 1960 to 1980 also saw the enactment of key environmental legislation related to air, water, and waste (Table 12.2).

When Rachel Carson's *Silent Spring* was published in 1962, the book alerted the United States and the world to the deadly effects of the industrial and agricultural chemicals humans were using. This book became a touchstone for initiating decades of research and activism on the impact of pollution and chemicals in the Earth's environment, including chemicals in materials that affect human health (see Chap. 8). While Rachel Carson authored the book, she was one of many scientists, bird watchers, farmers, journalists, activists, and other citizens whose concerns culminated in *Silent Spring*. Her book became one from a constellation of publications and events that profoundly affected public awareness. Landmark environmental regulations followed. Many of those regulations are still active today.

On April 22, 1970, Senator Gaylord Nelson led the organization of a nationwide "teach-in" about environmental issues. Twenty million citizens participated, accounting for nearly 10% of the US population in 1970. This teach-in is now known as the first Earth Day. It is partially responsible for creating the political climate that allowed the formation of the National Oceanic and Atmospheric Administration (NOAA) and the Environmental Protection Agency (EPA) on

Table 12.2 A brief timeline of events covered in this book: 1960–1979

1960–1979	
1962	*Silent Spring* was published by Rachel Carson
1963	Congress enacted the Clean Air Act of 1963 (see Chap. 8)
1964	Congress passed the 1964 Civil Rights Act
1965	Congress passed the Voting Rights Act of 1965
	Congress passed the Solid Waste Disposal Act (SWDA), allowing state control over waste (see Chap. 4)
1968	The first *Whole Earth Catalog* was published
1970	First "Earth Day" in the United States organized on April 22, 1970
	Founding of the National Oceanic and Atmospheric Administration (NOAA)
	Founding of the Environmental Protection Agency (EPA) by Richard Nixon's executive order (see Chap. 10)
	The Occupational Safety and Health Act was passed, limiting exposure to some toxic chemicals for workers (see Chap. 8)
1972	The UN Conference on the Human Environment held in Stockholm
	The Clean Water Act (CWA) replaced the FWPCA, now to be administered by the EPA, also establishing the National Pollutant Discharge Elimination System (see Chap. 6)
1973	The oil (energy) crisis began when Arab oil producers cut off exports to the United States [7]
1976	Resource Conservation and Recovery Act (RCRA) passed as an amendment to the Solid Waste Disposal Act (SWDA) enacted in 1965 (see Chap. 4)
	Congress enacted the Toxic Substances Control Act (TSCA) (see Chap. 8)
1977	The Department of Energy (DOE) is formed as a result of Jimmy Carter's executive order

October 3, 1970 [5]. Two years later, the UN Conference on the Human Environment (also known as the Stockholm Conference) was held in June 1972, marking the first major conference on international environmental issues [6].

In 1968, the Biologist Stewart Brand published the first *Whole Earth Catalog*. The catalog, among other things, brought the idea that human actions were radically changing the environment to millions of readers. Steve Jobs said the catalog was the "paperback prototype for Google," hinting at the breadth of its impact [8]. The second and third covers of the catalog were the "Earthrise" photo taken by Apollo 8 Astronaut Bill Anders, spreading the image across the United States.

The Historian Andy Kirk describes how the catalog marked a turning point in the way many perceived the relationship between humans and the environment:

> From the first sentence of the first issue, "We are as gods and might as well get good at it," Brand issued a bold call for a new kind of environmentalism. Decades before the term was coined, he argued that we were living in the Anthropocene, where human influences were altering conditions for life on Earth. Brand's optimistic vision of reconciling American technological know-how with environmentalism also appealed to broader audiences. With its call for readers to recognize their status as "gods," and its celebration of good tools and green technologies, the Whole Earth Catalog helped popularize the "appropriate technology" movement, which advocated for small-scale, decentralized and environmentally benign options. [8]

More recent publications have honored Brand's legacy with a similarly ambitious and global scope. They also mark a shift within the environmental movement toward an increased understanding and focus on the climate crisis. One example is Alex Steffen's *Worldchanging: A User's Guide for the 21st Century*, released in 2006 and updated in 2011. This book catalogs a compendium of solutions and ideas being employed around the world, with topics ranging from:

- Darfur Stoves, fuel-efficient cooking stoves deployed in the developing world
- Toronto's "Elevated Wetlands," a multilayered example of both bioremediation and transformative public art
- Tateni Home Care, South Africa, addresses three massive problems: poverty, unemployment, and AIDS by training unemployed youth to become paid home-care attendants so they can assist people with AIDS and other illnesses [9]

12.2.3 1980–1999

During the next two decades, commissions, protocols, regulations, and laws solidified changes introduced in earlier decades. At the same time, the building industry witnessed the development of sustainable initiatives and frameworks for establishing environmental metrics. This included voluntary standards and certifications which signified building owners' and manufacturers' commitment to sustainable initiatives (Table 12.3).

12.2.3.1 Corporate Social Responsibility and the Carpet Industry

In 1993, Paul Hawken published *The Ecology of Commerce: A Declaration of Sustainability*, with an urgent call to erase the imagined conflict between business success and sustainable environmental practices [12]. The following year, the founder of Interface Carpet, Ray Anderson, was asked by a customer "What's your company doing for the environment?" Soon after, Anderson read *The Ecology of Commerce* and was moved by its message [13]. He described the two events as culminating in a dramatic shift in perspective. This resulted in the redesign of Interface's product, manufacturing processes, and end of life disposal and recycling efforts.

"Doing well by doing good" became Ray Anderson's new business paradigm for success [14]. In his book *Mid-Course Correction*, he argued that this business model succeeded by earning customer loyalty by leading in sustainability, achieving

Table 12.3 A brief timeline of events covered in this book: 1980–1999

1980–1999	
1980	EnergyGuide program launched requiring a mandatory government label disclosing relative energy efficiency relying on data provided by manufacturers
1982	Brundtland Commission defined sustainable development for the first time (see Chap. 1)
1986	California passed the Safe Drinking Water and Toxic Enforcement Act, also referred to as Proposition 65 (see Chap. 8)
1987	Montreal Protocol adopted on September 16, 1987 (see Chap. 7)
1988	Bill McKibben published *The End of Nature* on global warming
1990	Congress passed the Pollution Prevention Act Congress enacted the Americans with Disabilities Act (ADA)
1991	McDonough and Braungart were commissioned to write the "Hannover Principles" (see Chap. 1)
1992	Energy Star government certification launched by the EPA and DOE (see Chap. 5) The Federal Trade Commission published definitions for recycled content in its Green Guides (see Chap. 4) The United Nations Framework Convention on Climate Change (UNFCCC) secretariat (UN Climate Change), the parent treaty of the Paris Agreement and the Kyoto Protocol, was established in 1992 to support a global response to climate change [10] The Green Label 2nd-party certification launched by The Carpet and Rug Institute
1993	The US Green Building Council (USGBC) was founded (see Chap. 1) The Forest Stewardship Council (FSC) was founded (see Chap. 4) Paul Hawken wrote *The Ecology of Commerce: A Declaration of Sustainability*
1994	John Elkington coined the triple bottom line (see Chap. 1) The concept of Environmental Product Declarations (EPDs) was proposed to ISO members for the first time, eventually resulting in the first EPD standard [11]
1995	The Sustainable Forestry Initiative (SFI) was founded (see Chap. 4) The first UN Climate Change Conference of the Parties (COP) was held in Berlin
1997	The Kyoto Protocol to the UNFCCC treaty was established The first version of ISO 14040: Environmental Management – *Life Cycle Assessment – Principles and Framework* was published

economic and resource efficiency, and setting a business example that forced the rest of the industry to catch up.

Anderson's message and leadership echoed the argument put forth by the management expert Peter Drucker in 1973 in his book *Management: Tasks, Responsibilities, Practices:*

> The fact is that in modern society there is no other leadership group but managers. If the managers of our major institutions, and especially of business, do not take responsibility for the common good, no one else can or will. [15]

Ray Anderson's initiatives at Interface Carpet influenced the carpet sector. The sector also responded to backlash in the early 1990s, where carpets, carpet pads, and carpet adhesives were implicated as the source of health issues from chemical off-gassing. As a result, The Carpet and Rug Institute (CRI) created the Green Label Plus certification, the first indoor air quality standard for carpet cushion [16]. By 2005, competition related to sustainability in the petroleum-dependent carpet industry provided an environment for the development of the first NSF/ANSI sustainable material standard, the NSF/ANSI 140 Sustainability Assessment for Carpet [17]. Many other product sectors followed the lead of the carpet industry and created their own NSF/ANSI standards for specific product types.

12.2.3.2 Green Chemistry

Current strategies to reduce toxicity and improve human health related to building materials focus on reduction or elimination of toxic chemicals as ingredients during manufacturing, as well as testing products before installation to ensure healthy air quality during the product's life (see Chap. 8). This approach of focusing on eliminating the source of pollution, rather than cleaning up polluting chemicals in the environment once they are waste, is a result of the ideas of green chemistry.

Dr. Paul Anastas is credited with developing the field of green chemistry in 1991 while Chief of the Industrial Chemistry Branch of the EPA [18]. Anastas collaborated with John Warner to develop a framework for preventing pollution when inventing new chemicals and materials rather than addressing them at the end of a process of product life. These ideas were presented in the *12 Principles of Green Chemistry* [19]. Warner co-founded two organizations dedicated to research and education on green chemistry in 2007: the Warner Babcock Institute for Green Chemistry and Beyond Benign.

While Anastas and Warner are credited with developing green chemistry as a field, pollution prevention was already understood as a necessary shift, evidenced by Congress passing the Pollution Prevention Act in 1990, an act that put emphasis on source reduction as a different, more desirable strategy than waste management of pollution control for reducing pollution [20]. The Act included energy, water, and resource efficiency as strategies for source reduction.

Green chemistry principles have evolved to keep pace with the increased number of toxic chemicals present in manufacturing today. This evolution includes the Six Classes Approach to chemicals discussed in Chap. 8. Regulations that ban chemicals at the source may only control a single industry or ban a single chemical rather than a class of chemicals, resulting in the replacement of one toxic chemical with a similar but differently named compound. This is also referred to as "regrettable substitutions."

Another figure in the constellation of green chemistry is Arlene Blum. She is a chemist responsible for contributing research that helped regulate two carcinogenic flame retardants used in children's sleepwear in the 1970s [21]. Decades later, she found that the same chemicals she had helped eliminate were being used as flame retardants in couches and other products. This experience inspired Blum to co-found the Green Science Policy Institute (GSP) in 2007 to address this challenge by educating policymakers and the public on how to better protect human health and the environment from toxic chemicals [22]. GSP later created the Six Classes Approach to educate policymakers and consumers on how to approach toxic chemicals in a way that avoids regrettable substitution.

In 2018, a more comprehensive ban of flame retardants continued the work of Blum's original research by prohibiting the entire class of chemicals. BuildingGreen's Paula Melton described the situation [23]:

> It's a familiar story: a toxic substance gets phased out, only to be replaced with a chemically similar one that has the same toxic properties. An amendment to existing California law aims to stop this pattern of "regrettable substitution" by effectively banning whole classes of flame retardants in upholstered furniture, mattresses, and children's products... By prohibiting entire classes of chemicals, the law has the unprecedented effect of banning future flame retardants that may be developed. "The State of California has found that flame retardant chemicals are not needed to provide fire safety," the original law reads.

As a result of the work of chemists and organizations like those described above, the elimination of toxic chemicals from building materials is expected as a core concept in most building material certifications and standards today.

12.2.4 2000–2020

The turn of the century saw an explosion of environmentally sustainable standards and certifications for both buildings and building materials (Table 12.4). Beginning in 2010, the focus began to shift from third-party certifications to often self-reported product disclosure or transparency labels such as Health Product Declarations (HPDs), Declare, and Environmental Product Declarations (EPDs), and JUST.

12.2.4.1 Sustainable Ingredients

In the 2000s, authors brought attention to the source of food and its associated impacts on environmental and human health, through publications such as Michael Pollan's *The Omnivore's Dilemma* in 2006. The desire to know an item's ingredients and where they came from continued into 2012, evidenced by Mark Bittman's "My Dream Food Label" in *The New York Times*, which included ratings for "Foodness" and "Welfare" alongside a food's nutrition [26]:

Table 12.4 A brief timeline of events covered in this book: 2000–2020

2000–2020	
2000	Leadership in Energy and Environmental Design (LEED) launched by the US Green Building Council (USGBC) (see Chap. 1) The GREENGUARD indoor air quality product standard and 3rd-party certification launched [24]
2001	The Stockholm Convention on Persistent Organic Pollutants was signed to eliminate or restrict the production and use of persistent organic pollutants (POPs), effective May 2004 [25]
2002	Tool for Reduction and Assessment of Chemicals and Other Environmental Impacts (TRACI) first released by the EPA (see Chap. 7)
2003	San Francisco set a goal to divert 100 percent of its waste from landfills by 2020 (see Chap. 4)
2005	The Cradle to Cradle (C2C) Certification was launched by MBDC (see Chap. 1)
2006	WaterSense label was launched by the EPA (see Chap. 6) The first version of the Living Building Challenge (LBC) was published (see Chap. 1) The 2030 Challenge was initiated by the non-profit Architecture 2030 Michael Pollan published *The Omnivore's Dilemma*
2007	NFS/ANSI 140 Sustainable Carpet Assessment Standard was published by NSF (later called NSF/ANSI 140 Sustainability Assessment for Carpet)
2008	Arlene Blum founded the Green Science Policy Institute, which would later author the Six Classes of Chemicals Approach (see Chap. 8)
2009	LBC launched its Equity Petal
2011	The HPD Working Group released the public draft of the HPD Open Standard Version 1.0 at GreenBuild 2011 (see Chap. 11) [26] "Material Rules" published in *Interiors and Sources* magazine
2012	The Declare Label was launched by the International Living Future Institute (ILFI) (see Chap. 11)
2013	The JUST label pilot program was launched by ILFI TRUE (Total Resource Use and Efficiency) was launched to certify zero-waste facilities (see Chap. 4)
2014	First version of the Natural Stone Sustainability Standard (NSC/ANSI 373) was published (see Chap. 4)
2015	The UN Paris Climate Agreement on Climate Change was adopted at the 21st Conference of the Parties (COP21) (see Chap. 7) The UN 2030 Agenda for Sustainable Development established the UN 17 Sustainable Development Goals (SDGs) (see Chap. 1) The Living Product Challenge was launched by ILFI (see Chap. 1)
2018	The 2030 Challenge for Embodied Carbon was launched by Architecture 2030, expanding the voluntary commitment from operational emissions to also include embodied emissions
2019	Greta Thunberg addressed the UN Climate Change Conference COP25 plenary in Madrid

FOODNESS – A measure of how close a product is to being real, unadulterated food. A piece of fruit gets five points, whereas fruit-flavored candy gets zero.
WELFARE – A measure of the impact of the food's production on the overall welfare of everything involved: laborers, animals, land, water, air, etc. This rating also accounts for carbon footprint and chemical (pesticide, for example) and drug (like antibiotic) residues.

This attention to ingredients also spread in the building industry, and interest in disclosure increased. In 2011, two architects from Perkins + Will, Robin Guenther and Peter Syrett, directly acknowledged Michael Pollan's book as an inspiration for their "Material Rules" [27]:

- If they won't tell you what's in it, you probably don't want what's in it.
- Consult your nose – if it stinks, don't use it.
- Just because almost anything can kill you doesn't mean building products should.
- If it starts as hazardous waste, you probably don't want it in your house.
- If it is cheap, it probably has hidden (externalized) costs.
- Question the generation of hazardous waste instead of where to use it in your building.
- Use materials made from substances you can imagine in their raw or natural state.
- Avoid materials that are pretending to be something they are not.
- Question materials that make health claims.
- Use carbohydrate-based materials when you can.
- Pay more; use less.
- Regard "space-age" materials with skepticism.

The first Material Rule, "If they won't tell you what's in it, you probably don't want what's in it" reflects the general interest in transparency and disclosure increasing during this period. The first version of the Health Product Declaration Open Standard, discussed in Chap. 11, was developed in 2010 by the Materials Research Collaborative led by the Healthy Building Network and BuildingGreen to help organize the increasing demands on manufacturers to present chemical ingredient lists for products [28]. The Declare Label launched 2 years later in 2012 by ILFI.

12.2.4.2 Social Accountability, Social Justice, and Equity

Throughout the twentieth century in the United States, labor laws and public health efforts from securing the right to collective bargaining in the 1930s to the widespread availability of vaccines have attempted to improve the health and lives of individuals. Acts, such as the 1964 Civil Rights Act, the 1965 Voting Rights Act, and the more recent 1990 Americans with Disabilities Act, have addressed issues of inequality and civil rights. In 2014, the campaign Black Lives Matter sprang out of a shooting in Ferguson, Missouri, a suburb of St. Louis, and grew to a global movement.

Despite the inclusion of inequality and social sustainability as key components of sustainable development as early as the Brundtland Commission, sustainability standards are still searching to address social justice and equity in a quantifiable way. Climate justice is one example of a movement aiming to fill that gap, bringing focus to the intersection of climate change and racial and economic inequality, since disadvantaged groups primarily bear the burden of climate change.

Development of quantifiable metrics for this area of concern within environmentally sustainable material standards and certifications has been limited. Early efforts include the founding of SA800 standard in 1989, the Social Equity Petal by the International Living Future Institute (ILFI) in 2009 and the JUST label in 2013. The JUST program aims to serve as a "nutrition label" for socially just and equitable organizations, reporting on the following indicators [29]:

- Gender and ethnic diversity
- Gender pay and pay-scale equity
- Living wage and full-time employment
- Occupational safety
- Employee benefits and worker happiness
- Local control and sourcing
- Responsible investing
- Charitable giving
- Community volunteering
- Transparency

12.3 From Individual to Collective Action

Environmental sustainability is often discussed in the context of individual actions and behaviors like taking public transit or biking, buying local, or recycling. Sustainability may evoke phrases such as "Reduce, Reuse, Recycle" or endearing wildlife peering from the cover of mail flyers from the World Wildlife Organization, the Sierra Club, or the National Audubon Society. These actions place the onus and focus on what individual behaviors are in relation to the environment.

Slowly, the focus is shifting from individual people, buildings, or organizations to broader collective action. For example, the 2030 Challenge initiated by the nonprofit organization Architecture 2030 challenges architecture firms to commit to lower the operational energy use of their building to reach net zero by 2030, authenticated through annual reporting. When the 2030 Challenge was initiated in 2006, many felt that clients should bear the burden of building energy use or that it was too difficult to achieve the targets set by the challenge. However, there are now nearly 1200 firms, organizations, and individuals that have adopted the 2030 Challenge [30]. In 2018, the 2030 Challenge was expanded to include embodied carbon. As of 2020, over 20 manufacturers and firms have signed on to the challenge, and a similar challenge has been established for structural engineering firms.

The UN Global Compact aims to create collective action among large companies, self-described as the "world's largest corporate sustainability initiative," with over 12,000 signatory companies in 160 countries as of 2020. The compact calls on companies to align their corporate sustainability goals with the SDGs and other universal principles on human rights, labor, environment, and anti-corruption.

As the author and activist Bill McKibben wrote in 2016 [31]:

> The problem is the word "I." By ourselves, there's not much we can do. Yes, my roof is covered with solar panels and I drive a plug-in car that draws its power from those panels, and yes our hot water is heated by the sun, and yes we eat low on the food chain and close to home. I'm glad we do all those things, and I think everyone should do them, and I no longer try to fool myself that they will solve climate change... North Americans are very used to thinking of themselves as individuals, but as individuals we are powerless to alter the trajectory of climate change in a meaningful manner.
>
> No, the right question is "What can we do to make a difference?"
>
> Because if individual action can't alter the momentum of global warming, movements may still do the trick. Movements are how people organize themselves to gain power—enough power, in this case, to perhaps overcome the financial might of the fossil fuel industry. Movements are what can put a price on carbon, force politicians to keep fossil fuel in the ground, demand subsidies so that solar panels go up on almost every roof, not just yours. Movements are what take 5 or 10 percent of people and make them decisive—because in a world where apathy rules, five or ten percent is an enormous number.

References

1. Hawken, P. (2007). *Blessed unrest - how the largest movement in the world came into being and why no one saw it coming*. New York: Penguin Group. Frontispiece text.
2. Fountain, H. (2017). Vikings Razed the forests. Can Iceland regrow them? *New York Times, Climate*. https://www.nytimes.com/interactive/2017/10/20/climate/iceland-trees-reforestation.html. Accessed 21 Sept 2020.
3. Mertz, L. (2017). *Origins of the interstate system*. United States Department of Transportation, Federal Highway System. https://www.fhwa.dot.gov/infrastructure/origin01.cfm. Accessed 21 Sept 2020.
4. United Nations. (2020). *History of the United Nations*. https://www.un.org/en/sections/history/history-united-nations/index.html. Accessed 21 Sept 2020.
5. United States Department of Commerce, National Oceanic and Atmospheric Administration, National Ocean Service (2019) *When was the first Earth day?* https://oceanservice.noaa.gov/facts/earth-day.html. Accessed 13 Sept 2020.
6. United Nations. (1972). *United Nations Conference on the Human Environment (Stockholm Conference)*. https://sustainabledevelopment.un.org/milestones/humanenvironment. Accessed 21 Sept 2020.
7. Myre, G. (2013). *The 1973 Arab oil embargo: The old rules no longer apply*. National Public Radio. https://www.npr.org/sections/parallels/2013/10/15/234771573/the-1973-arab-oil-embargo-the-old-rules-no-longer-apply. Accessed 21 Sept 2020.
8. Kirk, A. (2018). 50 years ago, the whole Earth catalog launched and reinvented the environmental movement. *Smithsonian Magazine*. https://www.smithsonianmag.com/innovation/50-years-ago-whole-earth-catalog-launched-reinvented-environmental-movement-180969682/. Accessed 7 Sept 2020.
9. Steffen, A. (Ed.). (2006). *World changing, a User's guide for the 21st century*. New York: Abrams.
10. United Nations Framework Convention on Climate Change. (2020). *About the secretariat*. https://unfccc.int/about-us/about-the-secretariat. Accessed 21 Sept 2020.

11. Bogeskär, M., Carter, A., Nevén, C., Nuij, R., Schmincke, E., & Stranddorf, H. (2002). Evaluation of environmental product declaration schemes. European Commission, DG Environment. https://ec.europa.eu/environment/ipp/pdf/epdstudy.pdf. Accessed 20 Sept 2020.
12. Hawken, P. (1993). *The ecology of commerce: A declaration of sustainability*. New York: HarperCollins.
13. Interface, Inc. (2020). *The interface story*. https://www.interface.com/US/en-US/sustainability/our-history-en_US#557997371. Accessed 13 Sept 2020.
14. Anderson, R. (1998). *Mid-course correction*. White River Junction: Chelsea Green Publishing Company.
15. Drucker, P. F. (1973). *Management: Tasks, responsibilities, practices* (p. 325). New York: Harper & Row.
16. BuildingGreen. (2011). *Green building product certifications: Getting what you need*. (p.71). Brattleboro, VT.
17. The Carpet and Rug Institute, Inc. (2020). *NSF 140: Environmentally-friendly products*. https://carpet-rug.org/nsf140/. Accessed 13 Sept 2020.
18. Warner Babcock Institute. (2020). *Green chemistry: A historical perspective*. https://www.warnerbabcock.com/green-chemistry/a-historical-perspective/. Accessed on 7 Sept 2020.
19. Anastas, P., & Warner, J. (1998). *Green chemistry: Theory and practice* (p. 30). New York: Oxford University Press.
20. United States Environmental Protection Agency. (2020). *Summary of the pollution prevention act*. https://www.epa.gov/laws-regulations/summary-pollution-prevention-act. Accessed 20 Sept 2020.
21. United States Consumer Product Safety Commission. (1977). *CPSC issues policy on Tris*. https://cpsc.gov/Newsroom/News-Releases/1978/CPSC-Issues-Policy-On-Tris. Accessed 22 Sept 2020.
22. Green Science Policy Institute. (2019). About. https://greensciencepolicy.org/about/. Accessed 22 Sept 2020.
23. Melton, P. (2018). *California law bans future flame retardants*. https://www.buildinggreen.com/newsbrief/california-law-bans-future-flame-retardants. Accessed 13 Sept 2020.
24. BuildingGreen. (2011). Green building product certifications: Getting what you need. p. 27.
25. Secretariat of the Stockholm Convention. (2019). *Overview*. http://www.pops.int/TheConvention/Overview/tabid/3351/Default.aspx. Accessed 21 Sept 2020.
26. Bittman, M. (2012). My dream nutrition label. New York Times sunday review: The opinion page. New York Times, 13 October. https://www.nytimes.com/2012/10/14/opinion/sunday/bittman-my-dream-food-label.html. Accessed 29 Sept 2020.
27. Guenther, R, & Syrett, P. (2011). *A (Pre)cautionary tale about buildings and food*. Digital Interiors and Sources Magazine. https://www.interiorsandsources.com/article-details/articleid/13005/title/a-pre-cautionary-tale-about-buildings-and-food. Accessed 13 Sept 2020.
28. HPD Collaborative. (2020). *About*. https://www.hpd-collaborative.org/about/. Accessed 13 Sept 2020.
29. International Living Future Institute. (2019). *Just. User manual*. https://living-future.org/wp-content/uploads/2019/01/19-0110-Just-Manual.pdf. Accessed 21 Sept 2020.
30. Architecture 2030. (2020). *Adopters*. https://architecture2030.org/2030_challenges/adopters/. Accessed 21 Sept 2020.
31. McKibben, B. (2016) *The question i get asked the most*. EcoWatch. https://www.ecowatch.com/bill-mckibben-climate-change-2041759425.html. Accessed 21 Sept 2020.

Index

H. Roth et al., *The Green Building Materials Manual*,
https://doi.org/10.1007/978-3-030-64888-6

W

Z

CPSIA information can be obtained
at www.ICGtesting.com
Printed in the USA
LVHW020844100521
686961LV00003B/227